ROADS
& Realities

ROADS
& Realities:

How to promote road contracting in developing countries

Paul Larcher & Derek Miles

Institute of Development Engineering
Loughborough University
2000

Institute of Development Engineering
Loughborough University
Leicestershire
LE11 3TU UK

Larcher P.A. and Miles, D.W.J. (2000)
Roads and Realities:
How to Promote Road Contracting in Developing Countries,
IDE, Loughborough University, UK.

ISBN Paperback 1 900930 03 X

Designed and produced at IDE
Cover photograph: Paul Larcher

Roads

I love roads:
The godesses that dwell
Far along invisible
Are my favourite gods.

Roads go on
While we forget, and are
Forgotten like a star
That shoots and is gone.

'Roads' in 'The Collected Poems of Edward Thomas'
by R.George Thomas (Ed). Faber and Faber, London, 1936 (p.263)

'One of the most obvious areas of Africa's decay is the infrastructure. The road is fundamental to the nation and yet it is in large parts in utter disrepair, for mile after mile. It tells us alot about the state of communications in Africa. It tells us alot about the African condition. It was Julius Nyerere, founder president of Tanzania, who once said that while the great powers are trying to get to the moon, we are trying to get to the village. Well the great powers have been to the moon and back, and are now communicating with the stars. In Africa, however, we are still trying to reach the village. And the village is getting even more remote, receding with worsening communications even further into the distance.'

Ali A. Mazrui
Professor of Political Science
University of Michigan

Foreword

Worldwide, road agencies are recognising the significant benefits of contracting out road construction and maintenance, which carries with it efficiencies driven by competition and the opportunity for contractors to develop and implement innovative techniques and materials. However, the transition from a direct labour-based organisation to that of a road manager, with the physical work being carried out by contract, is not an easy one and needs to be tackled on a number of fronts.

First and foremost the government needs to be clear on what it wants to achieve, so that it can set a sensible policy to be followed by road agencies. This is required to overcome any possible resistance within the agency. It is also necessary to develop policy in areas such as procurement standards and procedures to ensure that the letting and management of contracts is transparent and can be managed and scrutinised.

The road agency must prepare itself by developing its own contracting strategy that would involve a number of steps. This also needs to involve training for its own staff in the various forms of contract and the development and management of contracts. Road agencies may also have a role to play in the training of contractors and their staff in the application of various techniques and methods.

The strategy adopted in Spain for outsourcing of its road maintenance involves three distinct stages and could be considered as a possible model. They commenced with the direct hire of plant and labour under the direct supervision of their own staff. This served to train the contractor's people in road maintenance and start to develop the necessary capacity in the road maintenance industry. The second phase consisted of contracts based on a schedule of rates, where the contractor was issued with works orders and payment was based on tendered rates and measurement. The third and final stage will see the implementation of long term, five to ten year, performance- or outcome-based contracts in which the contractor is paid an amount each year to maintain a road network to a specified standard.

The period of time for the implementation of a contracting strategy in developing countries should not be underestimated. This could take at least five years and possibly longer in the case of performance- or outcome-based maintenance contracts. It is suggested that any road agency in a developing country that is considering contracting out of maintenance should seek out another road agency that has experience in the topic and enter into a twinning or other technology transfer arrangement with them.

The existence of a reliable source of funds and a reasonably consistent flow of funds is essential to successful contracting. Any breaks in funding levels for a period will result in a loss of the expertise that has been developed as people move to other areas or jobs. This is a good reason for the implementation of a road board of key stakeholders and the identification of a road funding source, preferably hypothecated based on vehicle registration or a fuel levy, so that road users can see how their funds are being spent.

Some opportunities may exist for contractors from developed countries to become involved with a local company. The contract could include conditions such as the level of employment of local people, a commitment to conduct training and some transitional arrangement whereby after a period of say 5 -10 years the external company could completely withdraw from the consortium.

Many road agencies have managed the transition from direct labour, or force account, to complete outsourcing by first implementing the purchaser/provider model within the agency and then privatising the provider arm, with the transfer of staff to the newly formed company. It is understood that a number of County Councils in the UK have taken this path.

One of the key ingredients in the effective management of a road network is the development of a capability in the private sector. The objective should be to provide a high level of competition and innovation so as to ensure that available road funds are spent as efficiently as possible, so that all countries can enjoy the economic and social benefits that come with a good road system.

Gary Norwell
Past Chairman
World Road Association (PIARC)
Committee on Road Management

Preface

The World Bank, in its *World Development Report 1994*, noted that infrastructure represents, if not the engine, then the "wheels" of economic activity. Thus it should be provided economically, and it should be properly maintained so as to serve the continuing needs of its users. The report estimated that, in Sub-Saharan Africa, almost $13 billion worth of roads - one-third of those built in the previous twenty years - had eroded due to lack of maintenance.[1] What went wrong? Most of the problems can be traced to the form of project planning and execution as well as inadequate management and unsuitable technology. There is growing agreement that the traditional public service way of managing roads is unlikely to deliver customer-focused efficient infrastructure in the 21st century.[2] This statement appears to apply equally to industrialised and developing countries.

While road agencies were starved of funds for routine and periodic maintenance, the development of local management skills in both local contractors and local road agencies was generally neglected. In short a market for the commissioning and execution of road maintenance hardly existed, and the smaller local contractors felt that they would be excluded from any opportunities that did arise. This book is based on the proposition that a form of market is needed which enables both groups to make an effective contribution to meeting the demand of the ultimate customers for access at reasonable costs. It examines the roles of the various stakeholders who are involved in the market, with a view to ensuring that road networks are managed as national assets in such a way as to enable the complex and varied needs of road users to be met economically and efficiently.

If the cost of access is to be reasonable, then the form of technology should draw substantially on local skills and resources. A feature of construction as an economic sector is its technological flexibility. Not only do the techniques used in construction tend, on the average, to be labour-intensive when compared with those employed in the rest of the industrial sector, but also there is a relatively wide range of technically feasible methods of carrying out a given construction project, some of which are very labour-intensive. The range of alternative techniques is probably widest in civil engineering, which is a relatively more important section of the construction industry in developing countries given the infrastructure requirements of industrialisation. In low- and middle-income countries, where the lack of employment opportunities is a common problem, labour-based techniques are clearly attractive providing that they are productive and competitive.

With productivity and competitiveness in mind, there are two key advantages in shifting from force account to contracting out road maintenance. Firstly, the road authority concentrates on its core tasks: monitoring the road network, identifying, programming, budgeting and inspecting the works. Secondly, the contracting out of road maintenance works helps to develop the local private sector and to promote employment.[3] For local entrepreneurs, a genuine and sustained market road maintenance opportunity would be likely to be welcomed, both as a direct source of workload and because road construction and maintenance provides a good entry into other types of public works and civil engineering projects.

A recent study on expanding labour-based methods for road works concluded that 'in many ways, small-scale contractors provide the only long term answer to getting value for money'.[4] The question of how to select or develop them is more problematical, but the effort is worth making if the outcome is a continuing national resource. This question is at the heart of this book, which was written to assist policy makers, programme designers and those responsible for the implementation of projects to support labour-based contracting for the roads sector in low income countries. In our view, it can only be tackled effectively within the context of the overall market and the role of the road agency as commissioner, designer, supervisor and general regulator. We believe that it is also necessary to bear in mind cross-cultural factors and a possible regional bias, since it cannot be assumed that a single model will be universally applicable.

The book is organised in six chapters. Chapter 1 provides a justification for interest in appropriate road technology (labour- and light equipment-based techniques) and the involvement of the local private sector in general and small-scale contractors in particular. Chapter 2 discusses the concept of a market for roads, including the role of contractors in roads and related markets, policies, strategies and the need for balanced regulation. Chapter 3 explains how to secure an enabling operational environment, including administration, management, procurement procedures, finance and the role of international technical co-operation. Chapter 4 is entitled purchasers and providers, and reviews the role of clients, contractors and market facilitators (consultants and NGOs). Chapter 5 is concerned with the design of contractor development projects, providing a model based on experience in Lesotho and discussing other approaches to contractor development, in order to suggest lessons for future projects. The final chapter examines ways of implementing change, including the need for supportive policies, the role of finance and funding, and methodologies and delivery systems to achieve sustainable performance improvements. It concludes with a section on project evaluation, based on ten short case studies, and suggestions for future research into the complex and neglected topic of international construction industry development.

References

[1]World Bank (1994) *World Development Report 1994: Infrastructure for Development*. Oxford University Press for the World Bank, New York.

[2]Dunlop, R.J. (1999) *Management structures need changing to drive the highway business in the 21st century*. Address to PIARC World Congress, Kuala Lumpur.

[3]Lantran, J.M. (1996) *Contracting out road maintenance activities: A world-wide trend*. ASIST Bulletin No 5, ILO/ASIST, Nairobi.

[4]Stock, E.A. and de Veen, J. (1996) *Expanding Labor-based Methods for Road Works in Africa*. World Bank Technical Paper No. 347. World Bank, Washington DC. p 45.

Acknowledgements

This book has been prepared as a component of the Management of Appropriate Road technology (MART) initiative, which is based on DFID Research Project R6238, and is led by the Construction Enterprise Unit, Institute of Development Engineering, Loughborough University, in collaboration with specialist consultants Intech Associates and I.T.Transport. The authors gratefully acknowledge the advice and assistance of their colleagues on the MART team, and the support of DFID in enabling this work to proceed.

The structure of the book draws upon a framework document[1] which was produced as collaborative effort between the MART initiative and the International Labour Organization (ILO), following a joint workshop held at Mazvikadei, Zimbabwe in 1995. Although we have departed from the framework in some of the following chapters, we wish to acknowledge the contribution made by all the workshop participants to its establishment.[2] We are, of course, solely responsible for the modifications to the original framework and for the opinions that are presented herein.

[1] Miles, DWJ (Editor). *Towards Guidelines for Labour-based Contracting: A Framework Document. MART Working Paper No.1*, Construction Enterprise Unit, Institute of Development Engineering, Loughborough University, 1996.

[2] C-A.Andersson, P.H.Bentall, A.Beusch, Ms B.Demby, R.Cadwallader, J.de Veen, H.Goldie-Scott, F.Hwekwete, W.Illi, P.Kanyugi, L.Karlsson, A. Kidanu, M.B.Kwesiga, A.Lehobo, S.Mazibuko, D.W.J.Miles, W. Musumba, S.Nyika, E.Opoku-Mensah, S.Otsuka, R.C.Petts, J.Runji, M.Shone, D.Stiedl, Ms E.Stock, T.Tessem, A.Twumasi-Boakye, J.Ward, R.Watermeyer and T.Wetteland.

List of acronyms

AGETIP	Agence d'exécution des travaux d'intérêt public contre le sous-emploi
ASIST	Advisory, Support, Information Services and Training (ILO project)
AT	Appropriate Technology
Austroads	The association of road transport and traffic authorities in Australia and New Zealand
DECO	Development of Construction Material Enterprises (ILO project)
DFID	Department for International Development
GNP	Gross National Product
GTZ	Deutsche Gesellschaft für Technische Zusammenarbeit
ICE	Institution of Civil Engineers (UK)
ILO	International Labour Organization
ITDG	Intermediate Technology Development Group
IYCB	Improve Your Construction Business (ILO project and training material)
LCU	Labour Construction Unit (Lesotho)
MART	Management of Appropriate Road Technology
MCR	Micro concrete roofing
NGO	Non-governmental organization
NICMAR	National Institute of Construction Management and Research (India)
ROCAR	Road Construction and Rehabilitation for labour-based contractors
ROMAR	Road Maintenance and Regravelling for labour-based contractors (ILO project, handbook and workbook)
TACETA	The Tanzania Civil Engineering Contractors' Association
TRL	Transport Research Laboratory (UK)
UNCHS	United Nations Centre for Human Settlements
UNDP	United Nations Development Programme

Contents

Chapter 1

Technology Plus Enterprise

This chapter attempts to answer several 'why?' questions; Why are good roads essential assets, which must be properly looked after and paid for?, Why use a labour-based technology?, Why involve the private sector?, Why encourage small-scale contracting as an alternative to direct labour or large contractors? It then moves on to the 'Wherefores'; explaining how these answers can be usefully combined in an approach which combines appropriate construction technology with enterprise development so that small-scale contractors are helped to execute road construction and maintenance using forms of technology that are appropriate to the special needs of developing countries.

1.1 Introduction

Access is a universal need, and transport facilities of various kinds are essential to the achievement of both economic development and social welfare. Although roads are only part of the solution to the transport problem in developing countries, a basic road network is vital to:

- enable goods and services to be delivered efficiently to their users;
- enable buses, taxis and private cars to move people from their homes to their places of work, to public and social facilities and to market places;
- allow general freedom of movement;
- enable national social services to be delivered throughout the country; and
- promote industry and trade.

Roads are assets

A good road system is therefore a desirable asset. Indeed in many developing countries it is essential rather than merely desirable, because roads are the sole realistic means of access for most individual travel and commercial trade. For example, it has been estimated that road transport accounts for 80-90 per cent of passenger and freight movement in Sub-Saharan Africa.[1] However, it is an asset that is costly to build and to maintain. In principle roads, like assets belonging to commercial enterprises, ought to be managed in a way that enables them to continue to serve their customers. The value of a 'road business' like the value of any other business, could well be measured by the current replacement cost of its fixed assets. The net current replacement cost of an item of plant or machinery is normally calculated by:[2]

(a) determining the asset's gross replacement cost; and

(b) reducing this amount by an accumulated depreciation provision based on:
    (i)   the gross current replacement cost;
    (ii) the proportion of the asset's useful economic life which has expired; and
    (iii) the depreciation method adopted for the asset.

Unfortunately this is easier said than done, even for items of plant and machinery, and 'attempts to calculate replacement cost run into the problems of changing technology, and the volume of work involved arising from the quantity and diversity of assets held by companies'.[3] For road networks it is even more difficult, since they cover vast areas and techniques for measuring replacement cost or value to their users are still subject to debate and development.

Road asset management

Despite the problems, there is a growing trend in a number of countries to find ways of regularly valuing roads as assets, with the implication that road agencies are expected to make the best use of limited resources so as to maintain, replace and/or preserve the assets for which they are responsible. They will also be held to account for any diminution in asset value due to neglect or poor management. To cope with these responsibilities a new discipline of road asset management has arisen (see box opposite).

Once roads are seen as a store of value which can yield continuing economic benefits to their users, the road agency can prepare realistic budgets based on calculated asset values, rather than go through the traditional process of 'asking for as much as possible, and hoping to be allocated just about enough to keep things going until the following year'. The central objectives of road

What is asset management?

Our decisions are sound, reasonable and appropriate when they are based on solid facts. All too often, however, decision making about resource allocation is based on anecdotal information and intuitive judgements. Asset management is a systematic process of maintaining, upgrading, and operating physical assets cost-effectively. It combines engineering principles with sound business practices and economic theory, and it provides tools to facilitate a more organised, logical approach to decision-making. Thus asset management provides a framework for handling both short- and long-range planning.

An asset management system lets decision makers have ready access to quantitative and qualitative data on an organisation's resources and the facility's current and future performance. It facilitates decision-making based on these data and on relevant "rules of thumb" and principles drawn from engineering, economics, accounting, risk management, and customer service to ensure efficient resource allocation and asset optimisation.

Asset Management: Advancing the State of the Art into the 21st Century through Public-Private Dialogue, U.S.Department of Transportation: Federal Highway Administration, 1997

asset management have been described by Austroads (the association of road transport and traffic authorities in Australia and New Zealand) as:

- to maximise the benefits that roads bring to the community throughout their life, or
- to minimise the long term cost of delivering a defined level of service, or
- ideally, to find an acceptable balance between these two competing aims.[4]

Austroads has defined eight separate elements of road asset management as follows:

- asset management strategy;
- physical treatments;
- management of use;

- asset features;
- asset condition;
- asset use;
- road system performance; and
- community benefits.[5]

Austroads notes that these elements 'interact strongly, and the challenge in asset management is to harmonise the contribution of different elements so that the combined effect is an enduring balance between costs and benefits'. Valuing road infrastructure assets is a relatively new skill, and Austroads has been one of the pioneers in developing and applying commercial accounting techniques to valuing road networks. The dilemma is essentially that:

> *changes in financial value of road assets must be explained in terms of non-financial factors such as the performance, service potential or consumption of the assets (the user's perspective), and road condition and expected future maintenance costs (the owner's perspective).[6]*

The primary task for those responsible for organising the road delivery system is to provide a satisfactory network for a wide range of users, and to ensure that it is repaired and maintained as necessary so that it is available whenever needed. There can also be related tasks for road agencies regulated in the broad national interest, such as poverty reduction and the promotion of productive employment opportunities.

Roads and jobs

One feature of the construction industry is its flexibility in terms of input requirements, which means that there is considerable scope for job creation through the promotion of labour-based techniques. This is particularly true for the earth and gravel roads which are typical in the rural areas of developing countries. Thus it is possible to decide deliberately to provide and maintain the road network in ways that contribute to national employment policies, providing that productivity is sufficient to make these techniques economic. The former consideration points to the scope for appropriate road technology, the latter to the need to make effective use of the private sector and the importance of applying effective management practices in both contracting and employing organisations.

Who pays?

Performing the tasks of road construction, rehabilitation and maintenance costs money, so who is to pay for the road network? Roads have traditionally been seen as a public service, to be provided and maintained from tax revenues (or from public sector borrowing that will be serviced and eventually repaid through taxation). However when finances are tight, and they usually are, roads are likely to be the first of the public services to be neglected and to fall into disrepair. Hence the attraction of finding a way to charge the user directly for the service as and when it is used. The problem is that charging users fairly for the benefits that they enjoy from the occasional use of roads is relatively difficult. Tolls can sometimes be a solution on major highways or river crossings, but they are difficult and costly to collect on an interconnected network of minor roads.

One approach to studying the problem is to split the overall market into component sub-markets consisting of different types of road with different predominant user types and technical characteristics. For example, there are important differences between urban and rural transport needs, while there are also technical differences between paved and unpaved roads. It is also possible to distinguish between the tasks of maintenance and rehabilitation of the existing network and new works required to extend it. Although roads can be treated as a part of the public infrastructure, they can also be viewed as commercial assets that are capable of earning a commercial return. The concept of 'commercialization' implies that roads should be brought into the marketplace, put on a fee-for-service basis and managed like any other business enterprise. This requires complementary reforms in four important areas:[7]

1. Creating ownership by involving road and transport users in management of roads to win public support for more road funding, to control potential monopoly power, and to constrain road spending to what users need and can afford.
2. Stabilising road financing by securing an adequate and stable flow of funds.
3. Clarifying responsibility by clearly establishing who is responsible for what.
4. Strengthening the management of roads by providing effective systems and procedures and strengthening managerial accountability.

This chapter provides a justification for interest in appropriate road technology (labour- and light equipment-based techniques) and the involvement of the local private sector in general and small-scale contractors in particular.

Entrepreneurial discovery

Road construction and maintenance is a component of the wider construction sector, and shares the characteristic of being subject to heavily cyclical demand. This is a problem for both road agencies and construction companies, since investment in staff training and development or the purchase of specialised equipment is difficult to justify when future workload is uncertain. Where Government contracts are concerned, the danger of payment delays is a further hazard. Nevertheless the private sector tends to react more quickly to changes in demand than do public sector-based direct labour organisations, and its flexibility enables it to offer a cheaper and more effective service than the typical direct labour organisation.

Why should this be? An explanation can be sought in what has been described as the theory of entrepreneurial discovery, in which new ways of doing things are discovered as a result of dynamic competition, made possible by an institutional framework which permits unimpeded entrepreneurial entry into both new and old markets. The success which capitalist market economies display is the result of a powerful tendency for less efficient, less imaginative courses of productive action, to be replaced by newly discovered superior ways of serving consumers – by producing better goods and/or by taking advantage of hitherto unknown, but available sources of resource supply.[8]

Essentially the private sector is often better able to identify new opportunities than the public sector, largely because the reward system is biased to encouraging a willingness to take a measured risk in the hope or expectation of a larger reward rather than punishing it. Even though road construction and maintenance may appear a routine activity well suited to bureaucratic control based on formal operating procedures, there are often better and cheaper sources of materials and other resources that may not be immediately obvious, while productivity can be enhanced by more imaginative management and leadership. In a competitive market, these initiatives will be reflected in keener tender prices and a more cost-effective service for road users. It should also make it easier for the road agency to focus on its essential task of enhancing the asset value of the road network for which it is responsible.

Problems with captive contractors

Agencies designing contractor development programmes are sometimes tempted to try to create a group of *captive* contractors, who will only be expected to work for that particular agency and who will have a narrow technical specialisation. This means that the new firms will be very vulnerable during periods when the agency is unable to keep up a flow of funded work and are consequently quite likely to fail, with the result that the investment in contractor training will be wasted. *Captive contractors* suffer from many of the disadvantages of a direct labour force (force account), since they are wholly reliant on a single employer and learn that political and other pressures may well be the most effective means of ensuring a continuing flow of work. The problem is made worse where the agency provides financial guarantees to support equipment acquisition (see Ghana case in Chapter 5), since it will be understood that the contractor will only be able to repay outstanding loans if the work continues to flow. Thus a shrewd contractor will appreciate that *not* repaying equipment loans is the most rewarding policy, since a high level of outstanding loans will mean that the agency cannot afford to stop awarding contracts.

Related infrastructure markets

Labour-based road construction and maintenance for a designated road agency can be combined with a number of related activities. This is better for the contractors, since they can diversify their client base and keep a more even work load. It is also better for the road agency, which will benefit from the broader experience of its contractors in applying more appropriate techniques. Clearly the alternative work areas should be genuinely *related*, or the contractor will become a 'jack of all trades, and master of none.' Fortunately there are a number of compatible work areas in which labour-based road contractors can engage productively and profitably. Within the technical speciality of highways, road works can be carried out for other clients, such as housing developers or factory owners. It can also be extended to other simple public works projects, including water and sanitation, or to infrastructure associated with general building projects. If the contractor owns one or more wheeled tractors, it will be possible to undertake a range of agricultural work or simply hire them out on an hourly or daily basis (with operator provided, to ensure that the equipment is not abused). The issue of related infrastructure markets for small contractors is examined in greater detail in Chapter 2.

1.2 Why labour-based?

In some industrial sectors, there is little scope for technology *choice*. Where there is a conventional way of getting things done that works well and provides a convenient and cost-effective service to clients in comparison with the alternatives, this attitude may well be correct. There are other sectors, including road construction and maintenance, where real alternatives exist and the correct solution is not so obvious. This section is intended to assist the decision-maker to appreciate the scope for choice, then to choose the right technology and finally to ensure that its application is not hindered by an counter-productive business environment. It may be helpful to start with a definition of technology itself [9] (see following box).

A definition of technology

Technology is the means by which we apply our understanding of the natural world to the solution of practical problems. It is a combination of "hardware" (buildings, plant and equipment) and "software" (skills, knowledge, experience - together with suitable organisational and institutional arrangements). But if technology is to be useful, it is not sufficient for it to be made available. It must also be applied and maintained, which implies a demand for a further input of a suitable range of human resources and skills. It is this latter input that is at the root of the difficulty in transferring technologies between different environments.

Source: Miles, 1982

Why equipment was introduced

For road construction and maintenance in low- and middle-income countries, the crucial technology choice lies between labour-based and equipment-based technologies. Originally, labour-based technologies were conventional throughout the world, and proved satisfactory for large and small construction projects, including most of the world's road and railway systems as well as massive projects like the pyramids or the Great Wall of China. Subsequently, plant and equipment were developed and introduced into construction operations for one or more of the following reasons:[10]

- to save time;
- to produce work of better quality;
- to save labour;
- to cut costs.

Although equipment-based methods *can* and *do* yield these benefits in some circumstances, the balance of economic advantage does not necessarily *always* lie with equipment-based methods. Furthermore, there are socio-economic issues that need to be considered by countries which would have to import relatively costly foreign equipment to replace relatively cheap local labour if they opt for equipment-based solutions. Saving labour is no achievement where it is the one resource which is in massive surplus. In fact, an obvious advantage of labour-based technologies is the scope for job creation.

What is less obvious is that it can also be a *cheaper* and *more effective* way of building and maintaining roads. Equipment, particularly specialised equipment, is expensive to operate and maintain in developing countries (particularly at times when currency exchange rates are unfavourable). Reliable operational data from comparable labour-based and equipment-based operations remains difficult to obtain, but there is general agreement that theoretical equipment output rates and operational lives are very rarely achieved in practice in most developing countries.

Why should this be? To achieve theoretical availability rates with specialist equipment, a back-up system is required to ensure regular maintenance and prompt repairs by competent mechanics, as well as readily-available spare parts. For the operator, even when equipment is available for work, acceptable productivity rates can only be achieved when clients provide a regular workload, since specialist equipment cannot be easily redeployed to other work when there is a sudden shortfall in road contracts.

A comparison

Table 1.1 offers a comparison between the relative costs of labour-based and equipment-based methods, according to different factors and situations. It draws upon a variety of World Bank and ILO experience, and is indicative of intrinsic cost differences rather than attempting to give figures. For each situation, there is an indication in columns three and four as to whether the cost of labour-based and equipment-based methods respectively are 'strongly higher', 'higher', 'neutral', 'lower' or 'strongly lower'. The final column gives the relative cost of labour-based methods, based on the net impact of assessments in the previous two columns.

As might be expected, where labour is in scarce supply, equipment-based methods clearly offer advantages. However, where the labour supply is abundant the reverse is true, particularly where low wage rates prevail and productivity is generally high. An abundant supply of equipment and spare parts clearly favours equipment-based methods, but this is a rare situation in developing countries, and labour-based and light equipment-based methods are much more reliable when there are likely to be delays in importing heavy equipment or specialist spare parts. Regarding type of works and location, labour-based methods are usually most competitive on small, scattered and remote projects, but are less likely to be competitive on projects where there is an abundance of longitudinal earthmoving (where equipment is likely to be more efficient and productive).

Table 1.1 Factors affecting the relative cost of labour-based and equipment-based methods

Factor	Situation	Cost of labour-based methods	Cost of equipment-based methods	Relative cost of labour-based methods
Labour	Scarce supply	↑↑	↑	↑
	Low unskilled wage rate	⇩⇩	⇩	⇩
	High productivity	⇩⇩	⇩	⇩
Equipment and spare parts	Abundant supply	⇩	⇩⇩	↑
	Scarce supply	↑	↑↑	⇩
Type of works	Mostly small or packaged into small sub-projects	↑	↑↑	⇩
Location	Scattered	↑	↑↑	⇩
	Remote	↑	↑↑	⇩
Design	Significant cut and fill, leading to substantial longitudinal earthmoving	↑↑	↑	↑

Key

↑↑	Costs greatly higher	⇩	Costs lower	−	Neutral
↑	Costs higher	⇩⇩	Costs substantially lower		

Source: Adapted from Stock, EA and de Veen,J. *Expanding Labor-based Methods for Road Works in Africa. World Bank Technical Paper No. 347. World Bank, Washington, 1996. page 9.*

Making investment pay

For contractors with limited capital, the key to survival is to limit investment in fixed assets so as to leave adequate working capital to cope with investment opportunities or delayed payment from clients. Where currencies are not stable and freely convertible, the risk associated with acquiring imported equipment with scarce foreign exchange further increases the risks inherent in the contracting business. While some investment in tools and equipment is essential, it is generally better to look for items which are *versatile* and *cheap to operate and maintain*. Versatility is vital because projects vary considerably and there will be periods when specialist equipment will be idle. Tractor-based equipment is particularly attractive, since tractors are ubiquitous in most countries and can thus be readily redeployed to other useful work when road contracts are not available, while access to spare parts and maintenance is likely to be cheaper and easier.

More employment

With rising population rates in most developing countries, coupled with high levels of unemployment and underemployment, the provision of additional employment opportunities is a general concern. As a result, labour-based methods may be specified in contract documents so that the contractor has no choice but to apply these techniques. In any event, the relative economic advantages of labour as against equipment offer a real potential cost advantage, *providing* the contractor has the supervisory capacity to manage a large labour force and labour laws are sufficiently flexible to allow the contractor to recruit casual labour to meet peaks in demand.

Job creation in target groups

In situations where the rapid creation of employment opportunities is a policy imperative, small labour-based contracts can be mobilised more rapidly than large complex schemes. It is also possible to create these opportunities within those communities most in need, and select and train target groups within those communities. Where the problem is urgent, the chosen technology may take the form of what Watermeyer (in the South African context) describes as *labour-intensive* (the use of as much labour as possible, by simply substituting labour for machines), as distinct from *labour-based* (also changing the technology employed so as to make it appropriate for manual construction methods).[11] There are dangers in losing sight of the need to achieve cost-effective solutions, as "make work" schemes can damage the reputation of labour-based technologies. In general, project work should be planned and executed economically and effectively, whether the technology is labour- or equipment-based, and the social benefits of job creation for target groups should be properly costed and explained to the client at the design stage.

Sustainable maintenance

Technology should be economic to maintain as well as economic at the stage of construction. Labour-based roads are maintenance-friendly, since local communities can learn to maintain them with simple tools and equipment rather than be dependent on central funding of road maintenance. The work can be carried out through community action, or by small local contractors and paid for by funds raised locally. Either way, there will be a strong possibility of the maintenance being actually carried out, rather than being included in a "wish list" for a budget prepared at a higher level in the government hierarchy.

Skilled staff are scarce

Conventional construction plant only achieves high levels of productivity when there is a substantial and continuing workload, which justifies the employment of skilled operators and maintenance personnel. Skilled staff of this kind are usually unwilling to work for small contractors in rural areas, so reliance upon heavy equipment-based technologies inhibits the capacity of small contractors to compete for available work in their home areas. Indeed operator training is economically wasteful unless the operators are willing to move with their equipment, since the skills learned are not likely to be transferable to other possible job opportunities in rural areas. In most developing countries, rural artisans can cope with operating and repairing tractors and simple intermediate equipment which is technically similar to agricultural equipment that is already in use.

Reduced capital repatriation

At a MART workshop in Accra, the Director of the Department of Feeder Roads noted that Ghana, like many developing countries, is characterised with the problem of deteriorating economic conditions, a crippling scarcity of foreign exchange and an abundant supply of cheap labour. Efforts have consequently been directed to developing and disseminating technologies which made more effective use of local resources (particularly human resources).[12] Labour-based technologies make comparatively little call on scarce foreign exchange. Even where new infrastructure is financed through 'soft' loans, the capital sum will eventually have to be repaid together with some form of interest. Conventional heavy construction plant is rarely manufactured in developing countries; its life is limited and there are continuing commitments to purchase specialist spare parts from abroad.

Reaching disadvantaged communities

Because labour-based infrastructure projects can be implemented on a small-scale, they are very suitable for execution by local communities. There are a number of local NGOs which are able to provide practical technical assistance to the communities and to indigenous entrepreneurs within them, so that they benefit both directly and indirectly from the construction and maintenance of new infrastructure (see following box for the example of Khuphuka in South Africa[13]).

Training linked to production: The case of Khuphuka

Khuphuka was established in 1991 as a voluntary association by a group of community leaders who were concerned at the lack of economic opportunities available to the majority of people in KwaZulu-Natal. Their objective was to set up an organisation that would, through **training linked to production**, provide people with an entry point to the economy, while strengthening community structures and promoting development. The priority target groups are 1) community groups who are engaged in development or about to engage in development, 2) unemployed young adults (especially women - at least 30 per cent), and 3) emerging and current entrepreneurs, such as micro manufacturers and building contractors. The Khuphuka concept is that the interests of these three groups are complementary, in that community groups could provide a market for the contractors, and the contractors could provide local employment opportunities. The catalyst for bringing this about is the range of Khuphuka training programmes based on the following precepts:

- Lack of exposure to formal education does not imply lack of intelligence.
- If an emergent contractor can do the job, then he (or she) can price it and plan it.
- Experience is a contractor's greatest asset.
- The majority of small-scale contractors would prefer to be independent operators, but circumstances may force them to become subcontractors.
- Training programmes should focus on confidence-building, sustainability and self-reliance.
- Training and development must go hand-in-hand.
- Time is money.

Instead of simply trying to provide short term (and probably unsustainable) employment opportunities, Khuphuka aims to create **employers**, who will in turn create employment. The prospective employers are members of local communities, who participate in the community development process as partners with Khuphuka and learn the skills of identifying and implementing project opportunities.

Source: Miles and Ward, 1998.

Stimulates local economy

Appropriate road technologies make maximum use of local resources; particularly local materials and local labour. Where resources are owned by local suppliers, the cash injection to the local economy increases local purchasing power and can therefore stimulate other kinds of economic activity.

Gender

The construction industry is, regardless of country, in many respects a conservative industry, and that is also reflected when it comes to the participation of women despite the fact that task rates for women are the same as those for men on the majority of manual tasks.[14] Studies on women's artisan training have shown how difficult it is for them to obtain conventional jobs with contractors or with public sector direct labour organisations and to retain them even if they manage to get occasional casual work. Locally-based projects using labour-intensive technologies have a relatively good record in providing jobs for women, as women artisans can be linked to a contractor working only within a limited geographical area.[15]. The fact that construction workers often have to move from place to place and live in temporary accommodation is a serious constraint on women's participation as they often have family commitments. Nevertheless there is an economic and social case for encouraging women's participation in the industry, since women are more likely to spend their earnings on the welfare of their families than on personal consumption. Small projects undertaken by small local contractors clearly offer better opportunities for women to participate on a regular basis.

Environmentally friendly

Labour-based projects encourage a more effective use of local human and other resources, so the impact on the environment compares favourably with equipment-intensive approaches which depend heavily on non-renewable resources for fuel, spare parts and eventual replacement. These considerations are obvious, but there are other factors which lead labour-based methods to be more environmentally friendly. For example, since labour-based road technologies are less competitive where bulk excavation is required, engineers experienced in these techniques will tend to keep road lines to natural ground contours, so the impact of the road on the physical environment is less severe than is likely to be the case where there is no cost penalty in extensive cut and fill.

1.3 Why the private sector?

Many of the early technical assistance initiatives to introduce labour-based technologies relied upon the direct employment of large numbers of casual or permanent labourers through some form of direct labour (force account) organisation. These initiatives proved that the technology was technically feasible, but productivity levels tended to drop when the early impetus was lost and activities became routine. It gradually became clear to both governments and donors/financing agencies that there was a need for competition and market disciplines if productivity was to be maintained at a level where labour-based technologies' performance could match its potential. In its 1994 World Development Report the World Bank drew attention to the global trend towards increasing private participation in infrastructure as a result of dissatisfaction with the performance of state-owned and state-operated infrastructure organisations.[16]

Efficiency and effectiveness

Competition is a driving force of excellence. It can lead to lower costs and improved services. Inefficiencies in State-owned enterprises arise not only from lack of competition but also from the absence of checks and balances inherent in private ownership; the pressures that shareholders and external directors exert on managers to improve efficiency, the pressures that capital markets exert on companies to allocate scarce resources economically and to operate within hard budget constraints, and the pressures that managers who are responsible to shareholders and directors exert on workers to improve productivity.[17]

As purchasers of contractors' services, road agencies are able to choose the most competitive bids for individual projects, and can then measure and compare performance by various contractors so as to encourage good performance in future awards. The contractors are committed to providing the goods or services specified in the contract at a given price, and can only secure a profit through exploiting one or more of five factors:[18]

- limiting overhead costs;
- increasing labour productivity;
- more effective site organisation;
- shrewd purchasing; and
- risk anticipation.

None of these are easy to achieve, and all require strong internal management disciplines. By leaving these disciplines to the contractors, the road agency can focus on its main responsibility of careful planning and scheduling of work. The outcome of this split of responsibilities should be a virtuous circle of steadily improving overall performance.

Flexibility

Most Governments have learned the dangers of excessive overheads through employing large numbers of staff on force account (direct labour). Financing for road projects is notoriously cyclical, and salaries and wages for direct labour staff have first call on funds when financing is reduced. This frequently means that funds for materials, transport and essential equipment are squeezed to the point where work cannot be carried out and the labour force is idle. By using the private sector, the road agency's capacity to carry out work can be expanded in line with growing needs when funds become available without incurring a continuing commitment to a large payroll. Small firms in particular are more able to manage their limited resources by making adjustments to meet fluctuations in demand.

Gender

It is frequently difficult for women to gain employment in a monolithic direct labour organisation, particularly when there are artificial barriers to employment such as years of formal education rather than practical knowledge and experience. Although some private sector employers will be equally conservative, the diversity of potential employers means that it is likely that some women will be able to secure employment. Providing that they then prove their worth as diligent employees, they will have demonstrated a competitive advantage which can lead to further opportunities.

Where employment is provided by the private sector, it is possible to provide short term stimuli to encourage the employment of women without the danger of a long term commitment. Examples can be specialist training courses, subsidies for day care facilities, help with the acquisition of tools or 'job auditions', whereby women offered to work on-site on a trial basis at no cost to the employer (with the proviso that payment would be made if a long-term placement was offered).[19] A further possibility is for women entrepreneurs to establish their own contracting firms, and compete for work directly in the market place.

Innovation

The pressures of competition, coupled with the diversity of individuals and groups within private construction firms, increases the possibility of devising new and more productive techniques. The idea of entrepreneurial discovery was introduced earlier in this chapter. It has been described as 'the alert becoming aware of what has been overlooked, as the essence of entrepreneur-ship consists in seeing through the fog created by the uncertainty of the future'.[20] In principle, public sector administrators are forced by the hierarchi-cal management structure in which they operate to be risk averse and are therefore reluctant to put too much effort into trying to see through this fog. Small entrepreneurs, on the other hand, tend to be alert to opportunities to reduce costs by alternative ways of doing things.

Cutting costs

Once a contractor is committed to a fixed price project, additional profits can only be secured by a continuing search for ways to cut costs by improving efficiency on all component activities. The pressures of competition are very real in the private sector, and contractors who have managed to cut costs will be able to offer more competitive bids in the following rounds of competitive tendering.

Better financial management and accountability

There is frequently little or no incentive for managers of force account operations in developing countries to reduce costs, as improvements in finan-cial performance rarely lead to direct benefits for those responsible and budgetary savings often lead to reduced allocations for the following year. Within the private sector, managers are expected to be continuously cost-conscious and improved performance will lead to greater responsibility and improved financial rewards.

1.4 Why small-scale contracting?

How small is small?

There are many ways of defining 'small' firms; according to the number of employees, turnover, size of contract for which the firm is permitted to bid, and so on. In general, this leads to a somewhat sterile debate and often to somewhat dubious statistics. For example a 'one man' or 'one woman' firm may take on a large contract and carry it out using a number of sub-contrac-tors. We accept that definitions are sometimes necessary, where benefits or contractual advantages are limited to a specific class of enterprise, but for the purpose of this book we assume that the reader is sufficiently experienced to recognize a small enterprise from a large one in his or her particular environ-ment. In low- and middle-income countries, where there is a dearth of medium-size construction enterprises[21] there is usually a marked dichotomy between a small number of large businesses and a large number of small firms.

Small firms create more employment
In its 1994 World Development Report the World Bank noted that technical change is making it possible for smaller scale operations to be economical - and implemented by private enterprises.[22] Over the past decade there has been a growing awareness of the importance of the social and economic roles of small business on the basis of four favourable characteristics:

- Small-scale enterprises are generally more labour-intensive than large firms and therefore generate more direct, and probably also more indirect, jobs per unit of invested capital.
- They provide productive outlets for the talents and energies of enterprising, independent people, many of whom would not fulfill their potential in large organisations.
- Small firms often flourish by serving limited markets that are not attractive to large companies.
- They supply dynamism and contribute to competition within the economy, enhance community stability and generally do less harm to the environment than large organisations.

The case for external action
It can be argued that it is sufficient to passively allow this gap in the market to be filled over time without external intervention, on the proposition that entrepreneurs are by definition self reliant, energetic and innovative, and do not generally need to be coddled by promotion programmes.[23] However, in many developing countries the transition from a centrally controlled economy to a market economy has been so rapid, and the need to create new and sustainable employment opportunities is so urgent, that there is a strong case for external action to accelerate the process of enterprise formation and to ensure that as many existing enterprises as possible survive and prosper.

The quiet conspiracy
There is therefore a growing interest in promoting small enterprise development activities within international technical cooperation projects. The strategy is well justified. It is the tendency to hope for quick results from limited and sporadic interventions that can lead to disappointments, since it is difficult to implement any form of small business assistance, and it is often even more difficult to devise an organisation which economically and effectively

co-ordinates all the different services which are necessary.[24] In order to overcome reluctance among funding agencies, there is frequently a quiet conspiracy among small enterprise enthusiasts to underestimate the need for adequate inputs to achieve sustainable small enterprise development.

The conspiracy is powerful because everyone is involved; donors and recipient governments would like to believe that a technical cooperation project with limited inputs can show dramatic results, while there is no shortage of consultants and specialists who are prepared to play their part by taking on short term projects which promise a spurious (and usually unmeasurable) self-sufficiency. Nor is this comforting delusion limited to small enterprise development. Many large projects (for readers in the United Kingdom, Concorde and Eurotunnel are obvious examples) take longer and cost more than everyone pretended to expect.

Problems with planned programmes

Since the cost of sustainable individual small enterprise development projects can be disproportionate to the benefits, there are advantages in adopting a programme approach, in which project interventions fit into a coherent long term strategy and resource inputs can be amortised over a number of projects. The problem is that picking winners among prospective programmes is no easier than picking winners among prospective projects. If the programme is to be successful it will need a common theme in order to capture the imagination of donors and prospective participants, but even the best ideas have to be tested and modified before they are ready for widespread application. This process can be impeded by the detailed planning that is required to secure a financial commitment to a prospective programme. Indeed it has been argued that planning by its very nature defines and preserves categories, while creativity, by its very nature, creates categories, or rearranges established ones, which implies that formal planning can neither provide creativity nor deal with it when it emerges by other means.[25] Thus there is much to be said for treating more projects as pilot projects, accepting that development is an inherently risky process, letting winners pick themselves and favouring projects based on an imaginative approach to problems which are known to be widespread.

Silent partners

The small enterprise sector is the 'silent partner' of the Appropriate Technology (AT) movement, since small businesses create the most jobs per unit of invested capital, and show the highest outputs per unit of capital and per unit of energy consumed. They also tend to spread wealth more evenly in a community than do large investment projects in which only the wealthy can participate.[26] They are therefore more likely to use labour-based methods than are large contracting firms, for whom the problems of managing very large labour forces may be too daunting.

Capital constraints
Small firms generally tend to adopt ways of doing things that conserve capital, if for no other reason than that they possess little of it themselves and experience great difficulty in borrowing it from others. In the construction sector, labour-based techniques are therefore intrinsically attractive since they minimise the need for expensive investment in specialised plant and equipment. A comparison of capital costs for labour-based and equipment-based techniques is included in Chapter 4.

Low overheads
Many small businesses are run from home, or even from stacks of papers in cardboard boxes in the boot of a car or the driving compartment of a pick-up truck. This can make for a somewhat chaotic system for maintaining written records, but it does mean that these firms are not burdened with heavy overheads. In fact the 'filing system' is often in the owner's head, and is based on a good retentive memory and a list of personal contacts. Such firms frequently perform fully to the client's satisfaction on simple repetitive jobs, and are able to offer keen prices since oncosts are largely profit.

Market flexibility
The construction industry is cyclical in most countries, but the changes from 'feast' to famine can be particularly unpredictable in low- and middle-income countries. This means that the more flexible the structure of the industry, the more efficiently will it respond to market demand. A network of small contractors relying mainly on labour-based techniques comes closest to meeting this requirement, since they are not reliant on specialist suppliers, do not require large stocks of specialist materials or spare parts, and are able to mobilise most of their labour and material requirements from local sources. Such firms possess a particular competitive advantage in bidding for small, dispersed projects in rural areas.

Inter-sectoral flexibility
Over-specialisation is a severe handicap for a business in a rapidly changing market. Labour-based construction techniques are not limited to road construction and maintenance, but are also applicable to most kinds of simple

infrastructure works. Thus firms using these techniques can cope more easily with a cyclical downturn in road demand by seeking out alternative markets in other sub-sectors. The opportunities in related markets are discussed in more detail in Chapter 2.

Easy entry

It is generally easier to start a small and simple business than to start a large and complex one, and many entrepreneurs find it easier to start in basic building or public works even if they aim to eventually move on into another sector. Thus encouraging a market for small scale contractors can also generally stimulate entrepreneurship in a country or region, thereby offering the prospect of longer term economic gains as the owners and managers of these businesses become more experienced and seek out new opportunities. Table 1.2 sets out four common groups who may be attracted by contract opportunities in labour-based road construction or maintenance.

Reducing risk

From a national point of view, a construction industry delivery structure based on large numbers of flexible and responsive small firms reduces the risk of economic problems and social unrest if there are rapid changes in the capacity to finance a road programme. These dangers are inevitably more acute where the delivery mechanism is based on large direct labour forces or large private contracting firms which do not have the capacity to react flexibly to market changes. Furthermore equipment-based techniques usually imply heavy for-

Table 1.2. Common entry groups as small road contractors	
Entry group	**Comments**
Force account road supervisors	Technically experienced, but may lack entrepreneurial drive and flexibility.
Building contractors	Lack knowledge of road techniques, but experienced in project-related business activities.
Managers/supervisors from larger firms	Technically and managerially experienced, but may have become over-dependent on external administrative support and systems.
University or college graduates	May not have experience to temper theoretical knowledge with practical business realities, particularly financial and commercial pressures.

eign exchange costs both for replacement equipment and spare parts, so that the system is inherently likely to collapse if the sources of foreign exchange dry up.

Using local resources

The proposed delivery structure makes maximum use of both local labour and local materials, and is therefore more likely to provide additional benefits and satisfaction to local communities. These communities increasingly expect to have a major say in infrastructure priorities in their own areas, and also expect that indirect as well as direct benefits should flow from the construction and maintenance process.

Poverty focus

Donors and commissioners of aid projects increasingly seek to target their inputs to benefit 'the poorest of the poor'. This is difficult to achieve with large 'integrated' aid programmes executed by large organisations through complex administrative structures. It is much more likely if execution is through small, locally-based firms which habitually make full use of local skills and resources. Providing a 'bottom up' approach to project negotiation is adopted, there will also be benefits in improved access and economic empowerment of disadvantaged communities.

1.5 Technology plus enterprise

We can now move on to the consequences of the conclusion that there is a strong case for promoting appropriate types of construction technology through small construction enterprises. The ideas of appropriate technology and enterprise are both important, and this section aims to define what we are trying to achieve by promoting this dual approach.

Creating employment

With the growing trend to downsizing and privatisation of direct labour organisations, many countries feel an urgent need to create new employment opportunities for those who have been displaced. This is more likely to be achieved with a large number of small enterprises serving diverse markets than with a small number of large conventional private companies that operate through more rigid employment policies and procedures.

Value for money

Small contractors are also often the only type of enterprise willing and able to work on small construction projects. At least they can potentially offer the best value for money to clients on small projects such as schools, rural health centres and village water supplies as well as low-cost roads, which can often have a major impact on the quality of life of poor people. Without a network of efficient small contractors these facilities are often difficult or expensive to provide.[27]

Promoting productivity

Efforts to increase employment should go hand in hand with efforts to increase productivity, so that the goods and services produced will be more abundant at a relatively lower cost. The cheapest and most effective way of raising productivity is by better management, which may include better site planning and the application of basic work study techniques. Contractors also often seek to set daily wage rates in a way that will provide incentives to encourage improved productivity. There are three payment systems which are commonly used by contractors.[28]

Day rates: Paying a worker a set daily wage regardless of output.

Task rates: An item of work (a task) is assigned to each worker which would take an average worker a day to complete. The worker can be paid and leave work when the task is complete (sometimes known as 'job and finish').

Piece rates: The work to be undertaken is divided into many small jobs (pieces) which may take a few hours each to complete. A worker can do as many tasks as he or she likes during a day and will be paid the agreed sum for each task.

Incentive schemes such as task or piece work systems can double or treble productivity in comparison with simple day rates. These systems may be prohibited in some countries on the grounds that they discriminate between workers according to ability and may be thought to encourage undesirable working practices. Nevertheless, there is little point in promoting the private sector without enabling contractors to improve their performance. Ultimately workers will only undertake as much work as they are willing or able to do for the pay offered. If governments wish to promote the use of the private sector with its perceived associated increase in efficiency, they must put in place legislation which encourages rather than constrains this increase in efficiency.

Improving working conditions

In cases where a pure incentive payment scheme is considered unacceptable, a combination of the three systems can be used. For example, workers could be paid a basic daily wage which offers about 50-70% of the usual prevailing daily wage. The remaining pay would be on a productivity basis, with the average worker able to earn the other 30-50% of a standard wage from undertaking a standard days work. Higher than standard productivity would be rewarded with a return in excess of the prevailing daily wage, but the worker has the stability of a basic wage irrespective of productivity.

While the main issue of concern to workers undertaking a job is the level of pay, there are a number of other issues which may require legislation by the government, to ensure a fair deal for the workers. There are certain working conditions which should be addressed on each site and may be stipulated in government legislation on working conditions. These include, but are not restricted to:

- access to potable water at the site for drinking by any employee;
- first aid facilities; and
- allowances for rest breaks.

Representation

Workers suffer from a similar problem to small contractors in that individually they do not have sufficient power and backing to be heard. It is necessary for both contractors and governments to recognise the benefits which can be derived from trade unions which can represent and speak for the common interests of all the workers. Trade unions may be able to assist governments and employees set realistic figures for minimum wages, advise on incentive payment schemes and negotiate acceptable working conditions. Contractors would find discussing issues with one or a few unions much simpler than attempting to meet and discuss issues directly with all their employees. Trade unions should not be used as an excuse by contractors for failing to address workers' issues on their sites, but as a mechanism for facilitating the resolution of disputes between employee and employer. In order for trade unions to be successful they must be recognised by governments and contractors. Legislation may be required to force contractors to allow their workers to join trade union organisations and take part in their activities.

Developing new capacity

New small firms can start with small contracts for routine maintenance, and then gradually undertake more difficult projects as their confidence and experience grow. The great advantage of such simple projects is that they require very little investment in plant and equipment, and are relatively easy to manage and control. They therefore enable potential new contractors to try out their skills without too much risk. Some will succeed and some will fail, but failure need not be catastrophic when the investment in capital assets is limited. For road agencies and road users, the great advantage of trying out large numbers of small firms is that the best among them will learn and grow so that there will be a continuing capacity to meet the evolving demand.

Skills transfer

The learning process will affect both the firms and their employees, who will acquire useful skills which will enable them to take advantage of other employment opportunities in their local areas. If the process is supported by a formal training programme, the technical and commercial aspects of the training can be accompanied by general vocational skills training to support broad community development.

Community participation

In addition to formal and informal small enterprises owned by individuals, work can be awarded to community contractors. In any event, it is likely that when work is 'sliced and packaged' into small parcels, small local contractors will enjoy a competitive advantage compared to larger firms based outside the area. As a result, relevant commercial and vocational skills will be retained within the local community and the community will develop a sense of ownership of their local road network.

Make whole industry more competitive

Over a period the growth of significant numbers of competitive small contractors should make the whole industry more competitive, resulting in better value for money for construction industry clients and the local communities that they serve. Some of these small firms will grow into medium-sized or even larger enterprises, thereby replacing existing large businesses which lose their competitive edge.

Equity

Using appropriate types of technology and commissioning work through small contracts tends to favour disadvantaged groups, since it is easier to modify approaches so as to encourage greater equity. Labour-based road activities tend to improve opportunities for women to participate, and part-time work is facilitated where expensive mechanical equipment does not result in heavy overheads. It is also possible to foster small enterprises which rely on disadvantaged groups, by financial preference at the bidding stage.

Build capacity for routine maintenance

By committing funds to specific projects demanded by local communities, there is less likelihood that funds and resources will be diverted than in the case of a large direct labour organisation. These organisations inevitably carry large overheads, and limited financial allocations are frequently mostly absorbed by commitments to wages for permanent employees and other fixed costs. The combination of appropriate technology plus local enterprise is likely to improve the capacity for effective routine maintenance.

References

[1]Pennant-Rea, R. and Heggie, I.G. (1995) 'Commercialising Africa's Roads', *Finance and Development*. Vol 32, No 4, pp 30-32.

[2]Sizer, John (1989) *An Insight into Management Accounting*. (3rd edition) Penguin Business, London, p 166.

[3]Sizer, John. (1989) ibid. p 167.

[4]Austroads (1995) *Road Asset Management in Australia: State of the Nation 1994/95*. An Austroads Working Paper, Austroads, Sydney, Australia.

[5]Austroads, (1994) *Road Asset Management Guidelines* Austroads, Sydney, Australia.

[6]Austroads (1995) *op cit*, p 5.

[7]Paget, G. and Wetteland, T. (1995) 'Management and Financing of Road Maintenance: Progress in Africa'. Paper to XXth World Road Congress, World Road Association (PIARC), Paris, p 25.

[8]Kirzner, I.M. (1997) *How Markets Work: Disequilibrium, Entrepreneurship and Discovery*. IEA Hobart Paper No.133. The Institute of Economic Affairs, London. p 31

[9]Miles, D.W.J. (1982) *Appropriate Technology for Rural Development. The ITDG Experience*. ITDG Occasional Paper 2. Intermediate Technology Publications, London.

[10]Barber, G. (1973) *Builders' plant and equipment*. Newnes-Butterworth, London, p 27.

[11]Watermeyer, R.B. (1993) *Labour-intensive, labour-based and community-based construction: There are differences*. Siviele Ingenieurswese, Julie, South Africa.

[12]MART (1996) 'Intermediate Equipment for Labour-based Roadworks: Accra, Ghana, 19 and 20 April 1996: Workshop Report'. *MART Working Paper No.5*, Institute of Development Engineering, Loughborough University.

[13]Miles, D. and Ward, J. (1998) 'Integrating Infrastructure and Small Enterprise Development within Low-income Communities: The Khuphuka Concept'. *MART Working Paper No. 12*, Institute of Development Engineering, Loughborough University.

[14]Jones, T. and Petts, R. (1991) 'Maintenance of Minor Roads using the Lengthman Contractor System'. Paper to 5th International Conference on Low Volume Roads, PTRC, USA.

[15]Andersson, C.A. (1991) *Women can build: Women's Participation in the Construction Industry in Sri Lanka*. Construction Information Paper CIP/3. ILO, Geneva.

[16]World Bank (1994) *World Development Report 1994*, Oxford University Press, New York.

[17]Prokopenko, J. (ed.) (1995) *Management for Privatization: Lessons from Industry and Public Service*, Management Development Series No 32, ILO, Geneva.

[18]Edmonds, G.A. and Miles, D.W.J. (1984) *Foundations for change: Aspects of the construction industry in developing countries*, Intermediate Technology Publications, London. p 30.

[19]Andersson, C.A. (1991) *op cit.* p 32.

[20]Kirzner, I.M. (1997) *op cit.*

[21]Edmonds, G.A. and Miles, D.W.J. (1984) *op cit.* p 28.

[22]World Bank (1994), *op cit.*

[23]Marsden, K. (1987) 'Creating the Right Environment for Small Firms', *Small Enterprise Development: Policies and Programmes*, International Labour Office, Geneva. p 41.

[24]Harper, M. and Soon, T.T. (1979) *Small Enterprises in Developing Countries: Case Studies and Conclusions*, Intermediate Technology Publications, London. p 90.

[25]Mintzberg, H. (1994) *The Rise and Fall of Strategic Planning*, Prentice Hall, New York. p 299.

[26]Miles, D.W.J. (1982) *op cit*, p28.

[27]Relf, C. (1987) *Guidelines for the Development of Small-Scale Construction Enterprises*, ILO, Geneva. p 7.

[28]Heap, A. (1987) *Improving Site Productivity in the Construction Industry*, ILO, Geneva. p 6.

Chapter 2

A Market for Roads

Markets of some sort come into being whenever a demand for some kind of product or service among one group of people can be met profitably by other groups of alternative suppliers. Sometimes markets are obvious, as when people compare prices for food or clothing in shops or market stalls. Sometimes they are hidden and hardly recognized, as has been the case of the market for roads while road networks were constructed and maintained by Government-employed direct labour and financed through taxation. Nevertheless, such markets do exist, and they are more easily recognised where road users are able to articulate their demands, independent road agencies are answerable to them and operations are carried out by contractors to meet specified standards.

2.1 Three groups

There are three main groups of stakeholders who have a legitimate interest in the business of providing and maintaining a road network; ***customers** (road*

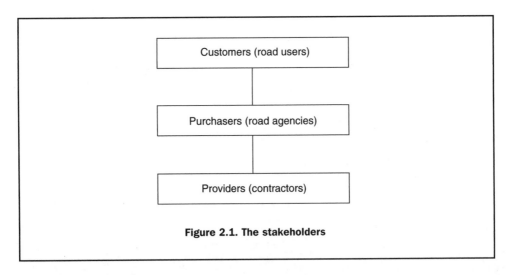

Figure 2.1. The stakeholders

users), **purchasers** *(road agencies)* and ***providers*** *of products and services (contractors).*

The effective functioning of such a market depends on a system which regulates the interaction of the various participating groups, recognising their separate roles and ensuring that their interactions achieve optimum results.

Separate roles

Each group has its own role, objectives and policies (usually fragmented and unco-ordinated in the cases of customers and contractors), but their interests can only be properly served if the other groups are satisfied with the working of the market in which they all participate. Traditionally in the roads sector, it has been common for a single public sector organisation to combine the roles of purchaser and provider, with the interests of customers represented by elected (or unelected) politicians who may (or may not) consult the views of their constituents. More recently there has been an increasingly distinct separation of roles, with road users seeking to exert influence through pressure groups (as do other interested groups, such as environmental lobbyists). There has also been a tendency to separate the purchaser/ provider roles, recognising that the functions of financing roads and providing them are very different. There is growing evidence that establishing and regulating a clear boundary between these functions can ensure that they are both carried out more effectively. The purchasing authority (road agency) can decide what is in the interest of its customers (road users) without being preoccupied with the convenience of providers, while providers have a clear contractual relationship and can focus on achieving targets in a cost-effective way. Let us now examine the characteristics of the three groups in turn.

Customers

Road users are the most varied group. Only rarely (as in the case of plantations or national park agencies and so on) is it likely that the type of use and the type of user can be accurately defined and controlled. Some use roads intensively and depend directly upon them for their livelihoods, such as those in the haulage industry. Others, such as village dwellers in low-income countries, may rarely make use of conventional roads and their needs may be catered for by simple tracks. This diversity makes the task of the purchasers and providers more difficult, since there is no directly identifiable target group for which a product or service can be designed.

Purchasers

Road agencies work on behalf of customers to identify and satisfy their demands at an optimal and generally acceptable level. They then mobilise the providers through the mechanism of performance-related contracts to achieve access at the agreed level. The task of the road agency includes fund raising to pay for the work of contractors, which may include direct user charges (such as tolls), indirect user charges (such as licenses) or a general road fund financed through local or national taxation.

Providers

Providers supply goods or services to the purchasers to enable them to meet the demands of the customers. In the construction industry, providers are normally *contractors*, who agree in advance to provide products and services of defined quality in defined quantities as set out in some type of formal *contract*. Thus contractors arise when there is a market need to be satisfied, including an ability and willingness to pay for the requirements. They can be categorised in various ways, but for most developing countries there is a significant division between foreign and indigenous businesses. Most of the foreign firms are either large general contractors or smaller specialist firms. Local firms may be large, medium or small, but it is the smaller firms that are most likely to be interested in smaller, dispersed projects that require the application of labour-based techniques.

Within the providers group, another division can be made between the formal and informal sectors, as indicated in Table 2.1 below.[1]

The division between formal and informal sectors is seldom completely clear cut, and many of the smaller formal sector contractors have transferred from the informal sector and only gradually abandon informal work practices. Part of the skill in designing contractor development projects lies in enabling and

Table 2.1. Indicators of the formal and informal sectors	
Indicators of the formal sector	*Indicators of the informal sector*
■ Licenses/permits required, relatively difficult to enter ■ Reliance on imported resources ■ Large scale operations ■ Corporate ownership ■ Capital-intensive (high foreign exchange component) ■ Formal skills, often expatriate ■ Market protected through tariffs, quotas and licenses	■ Easy to enter ■ Reliance on local resources ■ Small scale operations ■ Family ownership ■ Labour-intensive (local expense mostly wages) ■ Informal skills ■ Markets unregulated and competitive

Source: ILO, 1993

supporting this gradual transition, so as to help these businesses to cope with additional administrative requirements and inculcate sound business practices.

2.2 How markets work

Wages, profits and prices are determined by supply and demand working through the market. Markets can be regulated or distorted in order to achieve a variety of objectives, but they remain irresistable in that participants (potential buyers and sellers) will depend on market signals in terms of risk and reward in deciding whether to commission or offer products and services. The problem is that a market may exist between a road agency and the contractors who build and maintain roads, but usually no real market exists between road users and road providers. As a result, unless road agencies are extremely sensitive to the preferences of road users, 'since roads do not really belong to anyone, it is not surprising that they are used wastefully'.[2]

A perceptive paper by Lee illustrates the difference in linkages and pressures between private and public sector organisations.[3] Figure 2.2 is a simplified representation of the structure of the participants in a private market, where the primary transaction is between the firm, as controlled by its owners, and consumers. Owners seek profit by obtaining more in revenues from consumers than they pay to workers and suppliers. Relationships within the produc-

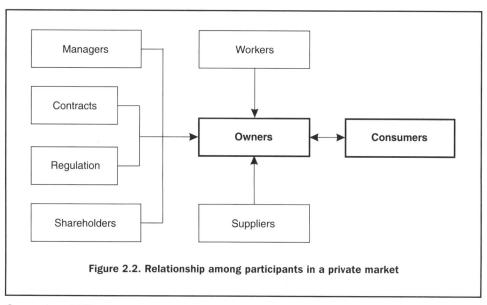

Figure 2.2. Relationship among participants in a private market

Source: Lee, 1995.

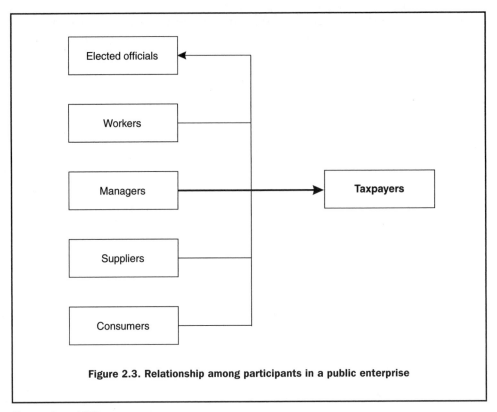

Figure 2.3. Relationship among participants in a public enterprise

Source: Lee, 1995.

tion side may be quite complex, but they are generally positive since all ultimately depend upon the firm's profitability (which in turn depends on providing an acceptable and competitive service to consumers). Contracts among participants and government regulation create an environment in which production, sales and income can take place. As Lee notes, 'the process often works badly, but the incentives are oriented toward reinforcing a common objective, and competition forces firms to constantly seek better ways to organise themselves'.

According to Lee, 'in a public enterprise - such as highways or transit - ownership resides in the taxpayers, but is effectively absent' in the sense that it is so dispersed that taxpayers are usually unable to meet as a group, define their priorities and ensure that these priorities are pursued by the road agency on their behalf. As represented in Figure 2.3, 'the self interest of all other participants is arrayed against the taxpayer, who provides only weak feedback (elected officials do lots of things which make the taxpayer unhappy and happy, if the taxpayer remembers them on the infrequent occasions of voting).

Which interest groups gain the most depends upon their relative political powers, but efforts to control costs or improve productivity amount to an uphill struggle for whomever undertakes them.'

Finance

In some countries, toll roads for primary national routes, motorways and major river crossings are commonly accepted, and the operators can identify a business proposition which is economically viable on the basis of user charges. In these cases there is a market for *road space*, and differential pricing can be used in which charges take account of higher demand (and consequent congestion) at specified times and/or places. Where no form of toll pricing exists, finance for roads comes in some form from general taxation. This means it is extracted from a wide variety of commercial organisations and individuals, some of whom are heavy road users whilst others derive only indirect benefits from improved opportunities for access.

Taxpayers do not pay for roads because they want to buy land and cover it with some form of all-weather surfacing. They pay for roads either unwillingly because they would otherwise be punished, or reluctantly because they want access and understand that otherwise it will not be provided. The difficulty is that they have to pay for general access for the country or region which is applicable for taxation purposes, while they personally are only likely to make occasional use of certain parts of the total network. As a result the unit price of using the highway is effectively *zero at the margin*. In this situation, which is also common for domestic water supply, the user has no incentive to economise, demand expands and always appears to be in excess of supply.[4]

Regulation

Regulation implies regulators, and detailed regulation depends for its success on the ability of one or a few people to judge what is best for consumers. Since regulation of utilities of various kinds is complex, it is inevitable that regulators exercise this judgement imperfectly. Accordingly, it has been argued that promoting competition is often a better approach than detailed regulation, since 'market competition is usually better partly because of its ability to surprise' and 'its results may well outstrip the daring of even the boldest regulator'.[5]

What form should the rules take? Transparency and clarity are often desirable but may entail inflexibility and bureaucracy, while they also depend to some extent on cultural differences (see following box).

Some national cultural views on rules and regulations

I heard a talk recently at which a Frenchman was illustrating some of the cultural differences between his business environment and that of the US. He quoted an extract from an American union agreement, which defined the order of seniority within the workplace. The principal criterion was the date at which someone joined the company. However, if two or more employees had been hired on the same day, then seniority was determined by the alphabetical order of their surnames. The rules went on to provide that if one of these employees subsequently changed his or her name, the order of seniority remained unaffected by the change.

The Oxford university audience laughed - until someone pointed out that these are exactly the rules that determine seniority within an Oxford common-room. Woe betide the newcomer, unaware of this, who sits in the senior fellow's place: still less the upstart who thinks he can usurp the senior fellow's place by changing his name from Zhirinovsky to Aardvark. The Frenchman found it amusing that these rules existed at all. But what was amusing to the English was not that the rules existed; what they thought funny was that someone had written them down.

The culture of rules and regulations varies between countries. The English assumption that decent people know instinctively what is right is so deep-rooted that we see no need for a formal constitution. For Americans, anything that is not explicitly prohibited is permitted. The French have elements of both systems, and neither: their joie de vivre rests on a curious combination of bureaucracy and anarchy.

John Kay, Financial Times

In principle regulations should aim for clarity, transparency and certainty. But this should not be at the cost of excessive bureaucracy and inflexibility. Good corporate governance is fundamentally about values and attitudes, so regulatory practices should help to develop positive values and attitudes.

2.3 What contractors want

There are advantages in enabling small contractors to compete for road construction and maintenance in low-income countries, since they tend to adopt technologies that make the most effective use of local skills and physical resources. The practices, problems and needs of small contractors vary from country to country, but there are some common problems. Table 2.2 sets out issues of concern to typical small contractors in Africa, developed at a MART international workshop in Zimbabwe.[6]

Table 2.2. Issues of concern to typical small contractors in Africa	
Issues	*Comment*
1. Definition of small contractors	■ What about the "self-made ......? ■ Some problems are country specific, so be cautious in offering generalised solutions ■ Classification is essential, and there may be a case for a special class for labour-based contractors
2. How to survive?	■ Start to operate ■ Registration and classification ■ Management ability ■ Have facilities ■ Have bank account ■ Marketing ability ■ Job continuity ■ Understanding the Government's budget cycle
3. How to get contracts?	■ Work closely with clients and contacts ■ Know what work is available (Media, Association, "Word of Mouth", relatives?) ■ Provide quality work within scheduled time and budget ■ A good record with claim resolution ■ Good tendering
4. Crucial problems	■ Late payment by client ■ Irregular payment by client ■ Lack of skills and difficulty in retaining staff ■ Lack of credit facilities ■ Unsuitable contract documents ■ Loose business partnerships ■ Compensation for late payment should be automatic, in the same way as liquidated damages is applied ■ Lack of job continuity ■ Poor contract administration by clients ■ Unrealistic specifications
5. Access to training	■ Prohibitive costs of training ■ Need to balance technical and business training ■ Lack of relevant courses ■ Lack of good trainers

Source: Miles, 1996

The two M's: Markets and Money

Most of the problems can be categorised within the two 'M's' that trouble most small businesses the world over: Markets and Money, although there is a recognition that performance could be improved with training in the third 'M' - Management. The issues of training and management development will be addressed in the second half of this book, but it is now timely to look more closely at the related issues of Markets and Money (which are discussed in sections 2.4 and 2.5 below).

In any market there are periods of 'feast and famine' but in the road market this cyclic fluctuation in the level of work can occur over two different periods. Firstly there can be a general rise and fall in the level of financial resources and hence work available in the road contracting sector as the importance of roads and transport links come in and out of political favour. Secondly, there is an annual change in the level of routine maintenance work due to the seasonal changes. Even when clients do make work available, bidding for contracts is often inhibited by oppressive prequalification and performance bond requirements, which exclude smaller firms.

Inexperienced small firms are often also structurally weak. Estimating skills are also generally deficient, due to a lack of knowledge of costs and an inability to evaluate risks. Even when a contract is awarded, small firms may need mobilisation advances and will be more vulnerable than larger firms if payments are not settled in accordance with the contract.

2.4 Related markets for small contractors

It is therefore essential that in order to be successful small scale road contractors should be able to turn to other contracting markets during periods of low workload in the road sector. The series of tables adapted from a publication of the Development Bank of Southern Africa entitled *Labour Based Opportunities in Construction*[7] show how labour-based small contractors could undertake a wide range of other infrastructure projects.

Agriculture

There is a wide range of work that the road contractor may be able to access in the agricultural sector. At a basic level it may be possible to hire tractors or trucks to farmers to assist them to transport their produce to markets in the local area. This service may also include labour to load and unload the produce. For small scale farmers it may be possible for the contractor to provide a transportation service for a group of farmers in one area which would take their combined crops to the central market, and avoid the need to deal through an intermediary.

If the contractor equipment fleet includes tractors, it is likely that these could be hired to farmers with the contractor's driver to plough fields or carry out other agricultural operations that require the use of a tractor. Large scale farms and plantation estates will have their own internal road network which will require maintenance. If a contractor has a proven record in rehabilitating and maintaining roads on the public network, he or she is likely to be also able to obtain contracts to maintain the private farm network. This type of work will be the most beneficial to the contractor as it is in the main business area and will make the most effective use of the existing labour force and stock of equipment.

Building

Many road contractors start out as building contractors. They should therefore be able to continue to undertake building projects when road sector work is unavailable. As the equipment and skills required for simple building projects

Building construction				
Activity	**Applicability**			**Notes**
	Good	**Average**	**Poor**	
1. Foundations ■ spread or strip ■ piled	✓		 ✓	If ground is not hard
2. Superstructure ■ brick/block ■ masonry ■ concrete ■ soil-cement	 ✓ ✓ ✓ ✓			
3. Frames ■ reinforced concrete ■ steel ■ timber	 ✓ ✓ ✓			If sections can be manoeuvred and placed by hand Assistance may be required with transportation
4. Roofs ■ timber truss ■ concrete ■ tiles ■ steel sheets	 ✓ ✓ ✓ ✓			Assistance may be required with transportation
5. Joinery	✓			
6. Internal Services	✓			
The design philosophy of the building will determine the level of labour based construction. In general low rise buildings are more suitable to labour based techniques.				

will be similar to those required for the construction of highway structures, contractors' labour forces and supervisors should be able to carry out the work to a suitable standard.

Civil works

Within the rural areas there is a variety of construction work that local road contractors may be able to undertake, such as irrigation schemes or hand pump surrounds and soakaways. Tasks of this kind require a modest range of hand tools and construction expertise which should be within the capabilities of small scale contractors.

Within the urban and peri-urban areas there is also a wide range of basic construction and maintenance work that road contractors may be able to undertake, such as:

1. Laying of water pipes and services
2. Erection of street lights
3. Construction of sewers and stormwater drains
4. Construction of communal latrines
5. Maintenance on urban infrastructure

For all of these tasks a contractor requires some simple hand tools and a small labour force with a supervisor, as well as a vehicle for moving modest amounts of construction materials. All these items and labour would normally be available to a road contractor. The majority of urban civil works would be commissioned by the urban councils where, as in the case of solid waste management, the contractor would have a head start in bidding for the work.

Electricity supply and telecommunications

Utilities, such as electricity supply and telecommunications, are now more open to the private sector. This change could result in new opportunities being made available to local contractors to undertake work previously carried out by direct labour forces. In the following two tables, activities 1-6 are common, and there may be opportunities for local contractors to carry out some work on electricity substations and cabling and maintenance of telecommunications systems. In some cases short training courses will be required in order to familiarise the prospective contractors with the requirements of the utility organisation.

Electricity supply				
Activity	*Applicability* Good	*Applicability* Average	*Applicability* Poor	*Notes*
1. Digging holes for pole erection	✓			Procurement and transport assistance may be required
2. Bush clearing and maintenance of area around lines	✓			
3. Trenching	✓			Ground conditions will affect the applicability of labour-based techniques
4. Backfilling trenches	✓			For the majority of backfilled trenches
5. Manhole construction	✓			Assistance may be required in materials procurement
6. Road crossings		✓		May require close technical monitoring and traffic handling measures
7. Substations ■ concreting ■ bricklaying ■ cleaning	✓ ✓ ✓			Close supervision may be required

Telecommunications				
Activity	*Applicability* Good	*Applicability* Average	*Applicability* Poor	*Notes*
1-5. As above	✓			
6. Road crossings		✓		May require close technical monitoring and traffic handling measures
7. Fitting overhead cables to existing pole routes	✓			A relatively unskilled operation, providing materials are supplied.
8. Maintenance of systems, clearing birds nests, weeds and other	✓			Small contract approach works well

Solid waste management

Within larger urban areas the collection and disposal of solid waste is co-ordinated by local councils. In many cases these councils contract out the collection and transportation of waste to disposal sites. Road contractors are well positioned to obtain this work, as they have the necessary labour and equipment. The contractors' gravel hauling equipment (tractors and trailers or tipper trucks) can also be utilised by each collection team to transport the waste to the disposal sites or transfer station. In tendering for waste collection, road contractors are well placed to compete as they have experience of how the councils operate their tendering system and preparing bids from their road contracts. They also have prior knowledge of the personnel to approach in the council, and the register will show the level of equipment that they hold.

In some areas councils need to construct large waste collection bins at the end of each street which are usually made from sand/cement blocks. When a council commences a solid waste collection service it is also necessary for them to construct waste transfer facilities where the waste can be sorted and sent to landfill sites or for recycling. These transfer stations require sorting bins, similar to the collection bins at the end of each street, and a few small offices. The small scale contractor is also well placed to undertake the construction of these facilities both in terms of experience and reputation within the council.

Solid waste management				
Activity	**Applicability**			**Notes**
	Good	*Average*	*Poor*	
1. Waste collection ■ domestic ■ commercial ■ industrial	✓ ✓	 ✓		Depends on the toxicity of the refuse and the availability of suitable equipment.
2. Provision of water collection points	✓			Small concrete/block work bins are required
3. Waste disposal ■ domestic ■ commercial ■ industrial	✓ ✓	 ✓		Depends on the material
There is also scope for job creation in the recycling of materials such as metal cans, glass, plastic and paper				

Water supply

There is a good range of potential activity in the vital task of water supply, as shown in the following table.

Water supply				
Activity	**Applicability**			**Notes**
	Good	**Average**	**Poor**	
1. Spring protection	✓			
2. Trenching	✓			The firmness of the ground will effect the applicability of labour based techniques
3. Backfilling trenches	✓			For the majority of backfilled trenches
4. Bedding for pipes	✓			Undertaken by labour even in equipment based work
5. Manhole construction	✓			Assistance may be required in materials procurement
6. Compaction ■ around sides of pipes ■ bulk fill above pipes	✓	✓		May require close technical monitoring to ensure required compaction achieved
7. Pipe laying ■ Small diameter ■ Large diameter	✓	✓		Problems may be experienced with handling
8. Reservoirs ■ small brick/ masonry ■ concrete ■ manufactured elements ■ plastic lined earth	✓ ✓	✓	✓	High material to labour cost ratio
9. Canals ■ excavation	✓			Assuming ground is not rock or very hard
10. Maintenance of water supply systems	✓			Small contract approach works well
The construction of sewerage systems follow a similar pattern to water supplies described in this table				

2.5 Contractors' access to credit

The main money problem facing small contractors, like all small businesses, is poor cash flow rather than (but sometimes, as well as) poor profitability. This is partly due to gaps in the flow of work and slow payment by clients, but also reflects a lack of access to and difficulty in obtaining credit. Compared to other small businesses, small construction companies have significant financial needs for short term working capital at certain stages of their projects, to cover materials purchases, staff wages and equipment purchase or hire costs. They may also need long term capital to cover the costs of expanding the business and financing the purchase, repair, maintenance and replacement of equipment.

The main problem facing contractors is their lack of collateral. They do not have significant premises where they work and most of their turnover is in the form of wages and materials. They also have very little fixed assets which can be used as loan guarantees. Contractors are often unable to provide a clear cash flow or profit forecast due to uncertainties such as weather conditions, worker productivity and materials availability. Banks therefore consider contractors and contracting a high risk business and are not keen to lend them money. If small contractors are able to obtain loans from the bank, a high level of interest is demanded to cover the perceived high risk associated with the loan. Banks in many developing countries are in general not set up to deal with small businesses and hence are unable to offer financial advice, have long cumbersome bureaucratic procedures for agreeing the size of loan requested and therefore have high administration costs compared with the size of the loan. All these problems result in contractors rarely obtaining money from established financial institutions, but relying on family and friends to loan the business money, based on ad hoc agreements.

A banker's viewpoint

A bank manager not only lends money belonging to the shareholders in the banking enterprise, but also money that belongs to investors that has been deposited with the bank for safe keeping. Managers are therefore expected to exercise caution when reviewing lending applications. Depositors may request their money back at any time and would not be pleased if the bank manager informed them that it had been lost on bad debts such as loans to unsuccessful contractors.

Before lending money to a contractor the banker will wish to review the applicant's past performance of operating a bank account. The banker is not solely interested in the account balance, but the frequency of deposits, if these deposits are regular or irregular, turnover in the account and the speed at which money is withdrawn after deposits. It is highly unlikely that a bank manager will lend to an unknown client who walks in to the bank and requests a loan.

If the potential client passes this stage the bank manager will then examine the actual loan request. He or she will wish to compare the amount of capital held by the contractor against the level of borrowing required, and the predicted cash flow for the project, to determine when and how funds are to be obtained and spent. The banker does not want the contractor to return to the bank the following week and request a further loan to cover new project costs, and will require a realistic cash flow forecast showing the maximum loan that the contractor will require during the course of the project. Once the maximum value of the loan is agreed, the banker will compare this figure against the contractors assets or capital. In general a loan is granted to supplement the contractors capital, so it is highly unlikely that a loan in excess of the contractors working capital would be approved.

If the contractor passes the first two criteria and the bank manager is still considering offering a loan, the final criteria must be satisfied; the loan collateral which the contractor can offer the bank. This security must have tangible value to the bank, i.e. it could be used to recover the cost of the loan if the contractor defaults. Items which can be used as security include, deeds to property, guarantees from other people or organisations and life insurance policies. In addition to this security it is highly likely that the banker will require payments due to the contractor to be paid directly into the bank.

Analysis of a banker's viewpoint on granting loans to contractors highlights a multitude of problems which small scale contractors face in obtaining loans. There are a number of different options which may be available to improve contractors' access to credit. These options each require the commitment from one or more organisations or institutions.

Mobilisation payments

Mobilisation payments are made by the client to the contractor at the beginning of a contract to provide working capital at the beginning of a job. Typically mobilisation payments are up to 15% of the contract value which represents about 2 months' income on a 12 month contract. During the remaining life of the contract the client will deduct a percentage of the contractor's monthly claim to cover the cost of the mobilisation payment. The advantage of this scheme is that the contractor will not have to obtain a large loan to commence work. The ultimate price of the contract may also be lower due to the reduced finance costs incurred by the contractor. The argument often cited by government officials against mobilisation payments is the risk that the contractor may not use the money for the intended project or even abscond with it. These risks can be minimised by careful selection of the contractor in the first place. The mobilisation payments can also be made directly to the contractor's bank which may also assist the contractor to obtain credit facilities from the bank.

Loan guarantee schemes

Contractors often find it difficult to supply adequate security for bonds or loan guarantees. In principle this might be overcome by the client or some separate government agency guaranteeing loans on behalf of the contractor. However, the risk of default must be kept in mind so there should be strict criteria for loan eligibility and careful screening of applicants, followed by careful monitoring and collection to prevent contractors treating the loan as a handout from the government. This can be assisted by initially only offering small loans and requiring the contractor to build a credibility rating through regular and prompt repayment in accordance with the conditions of the scheme. All payments to the contractor must be made through the bank where the loan is held. This scheme will require close cooperation between banks and the government, as contractors will require loans quickly once contacts have been awarded. There must also be mutual trust between the government and commercial banks, as each group is in effect controlling and responsible for a part of the other's money.

Banking code of practice

The construction industry is a very large potential source of business for the banking industry. Banks see contractors as high risk businesses partly due to the nature of their business, but also due to the way in which contractors often manage their finances. For example contractors may withdraw the majority of their monthly payment within days of paying it into the bank. In many cases there is a situation of both parties not understanding the problems and constraints that the other party has to contend with.

Banks could develop a code of practice for working with small contractors, which explains the services which they may offer and under what conditions. It may also be beneficial if the banks could produce a pamphlet aimed at small contractors, explaining the constraints under which banks can offer loans and how to apply for loans. Additional pamphlets could also indicate the range of services which banks may offer and how they can be beneficial to contractors. These pamphlets should explain why the banks have to insist on certain restrictions and selection criteria.

Establishment of construction banks

If the national banking industry remains reluctant to carry out business with contractors an option for the construction industry would be to set up a construction bank. This bank would provide contractors and consultants with working capital to undertake their projects. The bank, which would be a private sector-managed institution, would only offer loans to firms and organisations working in the construction sector. In order to commence, operation the construction bank would need a financial reserve. This reserve may be built up through loans and grants from the national government and donor agencies. Contractors who are working and receiving payments could also invest in the bank. Contractors' investment would entitle them to larger loans in the future. If the scheme is well established on a co-operative basis, the chances of default should be reduced as contractors are borrowing from each other and know each other as members of the same business community. It would also be considered socially unacceptable to default on payments, with peer pressure ensuring loans are repaid. Any contractor who defaults on the construction bank without a good or fair reason would no longer be entitled to apply for further loans.

2.6 Options for change

Where there is a transition from an integrated force account operation to a situation where a road agency is responsible for commissioning and control, while small contractors are responsible for execution, the responsibilities of the agency will have to be carefully defined. In particular, sufficient resources will be required to maintain an up-to-date road register and ensure careful planning of budgets and resources. Whenever possible, decision-making should be decentralised as close as possible to the user communities.

Contracting out of operational activities

There is a growing trend to outsourcing of operational activities. Where force account units are kept within the public sector, occasional 'market testing' by opening particular contracts to open or selected bidding is helpful in order to ensure that the unit provides a competitive service. It may also be helpful to introduce some form of incentive system so that the staff of the unit can benefit financially from achieving performance targets.

2.7 Policies and strategies

Initiating change

Government is, and is likely to remain, the predominant client for construction in most developing countries. The overriding factor in initiating change is therefore the commitment of the government at all levels. The government not only provides the majority of finance but also controls attitudes, policies, institutions and working laws. The foundation to initiating change must therefore be based on strong political stability and an eagerness or willingness to change.

The government must look at its whole political strategy as there will be constraints on both business development in general and the construction industry in particular (see Table 2.3). Each country is different: while there may be common features between different countries, analysis of their problems will identify specific problems and causes which will result in the proposal of unique menu of solutions. There are no standardised 'policy packages', each country or district will need to develop its own policy framework depending on the current economic, social and political environment. The objective is to provide a level playing field for all groups to compete on equal terms. While 'areas' may be targeted for support, careful consideration must be given to the effect on parties who do benefit from what may be support to a minority group.

A note of caution: the business world by its very nature has successes and failures. Whatever policy framework is initiated, all small businesses cannot be expected to succeed. The policy framework should not be expected to 'mother them all', but must have performance-based support policies and be

Table 2.3. Examples of policy issues for construction development	
Business development	*Construction industry development*
■ Poor banking systems	■ Long delays in payment
■ Lack of a legal system	■ Lack of consulting profession
■ Foreign exchange restrictions	■ Scarcity of trained workforce
■ Taxation system	■ Absence of design codes
■ Inefficient transportation systems	■ Lack of suitable contract procedures

realistic in allowing some unsuccessful businesses to fail. It must provide a free market and promote a competitive environment for the growth of the private sector. In short, it is more sensible to reward success than it is to reward failure.

The purchasers

The ultimate clients for public infrastructure are the population as a whole, but the immediate purchaser in most developing countries is usually a government ministry or public sector agency. This change in role for the public sector will require flexibility and an attitudinal change in order to adapt to the client role. They must either develop working practices and a capacity which will allow them to develop and oversee the projects that they are paying for, or they must assist with the development of consulting practices who are able to act on their behalf. In most cases, it is unlikely that that government departments could develop a sufficient combination of technical and entrepreneurial skills to undertake the client's responsibilities, design of the project and supervision of its implementation. The development of a consulting capacity will also provide an essential resource for private investors and developers to develop their own infrastructure.

The providers

Physical construction activities will be undertaken by contractors. However, this term includes a wide range of organisations, including large international multi-disciplinary companies employing thousands of staff, large national firms who employ local staff, and small contracting firms owned by one person operating in a small region with a handful of staff. Some of the local small contracting firms may not own any equipment, but act as sub-contractors providing a labour force to larger firms. Regardless of their size, all contractors must have access to requisite technical skills such as placing concrete, erecting formwork and bricklaying. However, as their size increases they need increasing management ability to cope with competitive bidding, risk management, financial planning and labour management.

Many local contractors, particularly small local contractors, lack the management skills (as well as the financial resources) to undertake large jobs. It may therefore be beneficial for local contractors to work in joint ventures or as sub-contractors to larger national or international contractors in order to develop their capacity to undertake larger projects.

Finally the providers could be the community themselves, through community-based organisations or self help schemes. The government should not overlook the potential and ability of these groups to implement projects. However, these groups may often lack some of the technical skills in addition to the management skills necessary to undertake construction work. Although they can form highly motivated 'contractors', careful and detailed supervision is often required to ensure a quality product.

Using local resources

Small scale contractors by their operating nature are locally based. They will make use of as many locally available resources as possible. These local resources include not only materials and equipment but also labour. The policy framework must therefore take into account the requirements of small businesses to use these local resources and introduce policies to facilitate this.

Contractors working in rural areas face logistical difficulties in obtaining imported or centrally manufactured materials. They prefer to use materials which are readily available in their area. Unfortunately these materials are often not listed in the construction codes. If they are to be used national construction codes need to be developed which would facilitate the use of local materials such as timber, masonry and locally manufactured blocks.

Rural contractors can also face problems in procurement, due to import restrictions and taxes, and further operational problems with equipment due to poor support services. Ways of mitigating these effects will be dealt with elsewhere. However, these problems can be circumvented by having positive policies towards the use of locally manufactured equipment and labour. Contracts and specifications could allow contractors to choose between labour or equipment based techniques without a penalty for either. Regardless of the approach that a contractor adopts, it should be possible to make use of locally manufactured equipment such as dead-weight rollers. The use of local resources will not only allow small scale contractors to be more competitive but will also offer a net economic advantage to the community as a whole. Firstly, the use of local labour will ensure that the local community has financial resources which will promote other tertiary businesses and secondly the use of locally available resources will promote primary and secondary businesses to supply the material needs of the construction community.

Decentralisation

There is currently a trend to decentralise construction and maintenance due to its perceived efficiency derived from a knowledge of the local working conditions. A policy of decentralisation enables local staff to rapidly make decisions which are appropriate to the local environment rather than waiting for queries to be sent to head office and then waiting for an uninformed decision to the sent back. While it is usually advantageous to decide policy centrally and maintain a funding body, decentralisation of planning, management and day-to-day finance decisions will have a beneficial effect on the development of the local contracting sector because:

- contracts are likely to be packaged into smaller units which favour small local contractors;
- work will be advertised in the local area rather than just in the major town or city; and
- the contractor will not need to travel to the large towns to meet the client for decisions or to secure payment, thus increasing efficiency and the level of supervision.

When deciding on the level and degree of decentralisation, the amount of work being undertaken and the capacity of the local and regional government to manage the construction process should be taken into account. Essentially the decision making powers should be delegated to the lowest district or regional level bodies who are still capable of making educated and appropriate management and financial decisions.

Attitudes, skills and knowledge (ASK)

It is essential that government officials are motivated to support new policies actively rather than just pay lip service to them. This will mean developing their *skills* and *knowledge* in these areas and also, most importantly, using participative workshop sessions to encourage a positive *attitude* to unfamiliar issues such as private sector implementation, use of more appropriate technologies and contractor development. These three aspects of training are brought together in the acronym ASK (Attitudes-Skills-Knowledge), and professional development courses should be arranged in all three ASK skills to ensure that they accept and support new policies.

Training courses can be arranged to assist with the development of the second two traits, but even participative workshops may not be sufficient to change attitudes. It is worth remembering the old rule:

Tell me	I'll forget
Show me	I'll remember
Let me try	I'll understand

Experience from past technical co-operation projects of the *show me* approach have included study tours for policy makers and senior officials, to other countries who have been undertaking road projects and/or contractor development programmes. These tours aim to introduce them to the principles and benefits of the existing projects so that they can return to their own departments for the *let me try* stage with fresh ideas and a 'new outlook' on the ideas.

A study tour to Kenya

A chief engineer had been attempting to convince his minister that the way forward in road maintenance was to contract the work out to small scale contractors. The Minister of Works was not interested in these ideas until he saw the scheme working in practice, outside his own country. While visiting a site which was obviously operating efficiently he turned to his chief engineer and asked "why are we not doing this already ?"

The ASK process takes a long time to achieve, both for government officials and contractors, since it is a professional development process and requires a skill to be taught and then practised in order for the participant to become knowledgeable and proficient in that skill. The knowledge can be taught on a training course, but it must then be practised on the job to develop the skill to implement it. Educational and professional development should therefore be seen as a long term process, rather than a short term training input at the beginning of a project.

Community involvement
It was mentioned above that the primary goal for creating an enabling environment was a positive policy and willingness to change from the government. Of equal importance is the willingness of the community to be the recipient of change and the 'owner' of road and other construction projects. They must feel that they have control of the decision making process. This involvement will affect the level of decentralisation of the decision making powers of the government. Both the government and the community must feel that they have confidence in the management ability of the organisation or office which is assigned to oversee a project. An obvious advantage of a labour-based construction and maintenance approach is that it is possible to promote the

participation of the local community who are likely to have access to the majority of the tools required for road works.

Where the community takes on the role as contractor through a community based organisation, simple but formal agreements or 'community contracts' should be made between the community and the authority which clearly set out the roles and responsibilities of each party in relation to construction, funding and maintenance arrangements. It will be necessary to allow the community to make important decisions supported with the technical guidance of the public authority.

Planning and co-ordination

Construction is a necessary activity in all government sectors; infrastructure, education, health, transportation, defence to name a few. However, it is often not represented by any single government department (except perhaps a Ministry of Works, which sees its role as primarily short term procurement rather than strategic development of the industry upon which it depends), which results in neglect of the industry as an economic sector. In many countries road construction and maintenance is overseen by a road department which is usually part of the Ministry of Works. By chance, as the government accounts for about 80% of the construction turnover in many countries, this ministry also oversees the majority of other construction activities. This situation highlights the need for the departments within the ministry to co-ordinate and plan together their work. One of the major problems that all small scale contractors face is the lack of work continuity. The construction industry is by nature cyclical, but by better co-ordination of work between government departments, and within the ministry of works itself, a steadier level of public sector construction demand could often be achieved. This steady level of construction demand does not include large projects such a dams, but projects which may be undertaken by small contractors, for example schools, health centres and minor infrastructure work.

In addition to co-ordination between ministries responsible for construction it is vital that the ministries of planning and finance are included in the construction planning process. By achieving a steady level of public sector construction demand, the ministry of finance should have an easier task in budgeting for the meeting the payment demands of the other ministries. Where a large amount of co-ordination is required between separate ministries, the ministry of planning could act as a central co-ordinator to achieve a balanced demand. It may be argued that co-ordinating work between road and other departments is not worthwhile as the construction work is completely different. In reality contractors will often undertake different work in order to maintain a steady work load and are likely to have different supervisors to oversee different work projects. For example, building contractors would be able to undertake construction and repair of highway structures such as small bridges and culverts and road contractors would equally be able to undertake water and sewage pipe laying contracts.

2.8 Regulation: The key issues

Government, as the main client in most developing countries, has an over-whelming influence on the market. Different parts of the government also have differing interests in the projects to maintain and construct new roads and the development of an indigenous construction capability. They should therefore act and legislate carefully in order to maintain a suitable balance between representing the interests of consumers of the product (the community), the financial stake invested and the interests of the ministries to be seen to be undertaking work.

To be effective, any new policies must be viewed a open and fair by both the public and private sectors. It is unlikely that the public sector capability would be dissolved immediately in favour of the private sector. In such cases, measures must be implemented which allow the public and private sectors to compete on equal terms, or work should be allocated between the two groups on strict guidelines and criteria. In many countries there may be the potential for cartels, monopolisation and corruption. Contracts must be awarded and contractors selected under a scheme which is open and fair, preventing accusations that foul play has occurred. Measures that can be implemented include:

- Public opening of tender proposals.
- Selection of contractors by a tender board rather than individuals.
- Selection of contractors according to an agreed ranking framework rather than subjective decisions.
- An observer from the contractors' association being invited to sit on the tender board.
- A limit to the share of total workload that can be awarded to a single contractor.

Risk transfer

In any contract there are inherent risks, and these are especially significant in construction. The government may be tempted to pass all these risks onto the contractor, with the perception that the client (government) will thereby get a better deal. In reality risk transfer has to be paid for, and the contractor will increase the bid price to take into account of these risks (plus an allowance for overheads and profit). There are then three possible outcomes:

- The contractor makes a large profit if the risks do not occur.
- The contractor absorbs the loss from some of the risks from the overall risk allowance.
- The contractor makes a large loss if all the risks occur, or in the worst case becomes bankrupt and is unable to complete the work.

Clearly outcomes one and three are not beneficial to the client, especially if the contractor becomes bankrupt and the client has to make further payments to an additional contractor to complete the work. In addition, for outcome two the client is likely to have had to pay a high contract price for the contractor to absorb a number of losses over a run of contracts. A contractor who is burdened with all the risks will always be looking to reduce costs in order to provide a 'financial insurance' against any of the risks occurring. The general principle is that risks should be borne by the party best able to assess and cope with them. For small projects undertaken by local contractors, it will usually be the client who is best able to bear the risk (and who will be compensated by benefiting from keener bids). Policy makers should remember that clients and customers are not the only stakeholders in the roads market. The contractors and their workers also have a strong interest in achieving a workable, equitable and reliable market.

A fair deal for contractors

Contractors run a construction business for one reason - to make a profit. However, to prosper over a reasonable period they must work with the government and other client organisations to achieve a workable, equitable and reliable market. The problem most contractors face is that individually they are too small and powerless to influence government decisions. The solution to this lack of individual influence is to establish a contractors' association which can represent the interests of all contractors, acting as one voice and providing a link between the government and many small and large scale contractors. A contractors' association can promote the interests of the construction industry, discuss its problems with the government, provide technical advice to contractors and act as a common negotiating body. Sensible clients will appreciate the advantage of discussing issues with contractors by talking to one body rather than attempting a dialogue with a multitude of individual firms.

For contractors' associations to be successful they must be recognised and receive mutual regard and respect from both contractors and government organisations. It is important from the contractors' viewpoint that a contractors' association is seen to be independent from the government. While recognition from government is essential, it is necessary for the organisation to be financed, staffed and managed autonomously. In most developing countries, contractors' associations are likely to require financial assistance to commence their work. Some money may be raised by members' subscriptions, which should meet the operating costs of the organisation. Government and/or donor contributions could be used to provide seed capital to establish services such as loan schemes and financial guarantees. By offering financial assistance to small scale contractors the association will raise its credibility, providing that the scheme is operated in a demonstrably objective and fair manner.

Membership of the contractors' association can act as a guarantee that the contractor is a legitimate, properly qualified company. While contractors should not be forced to join, non-members will not get the support and accreditation they require. Clients will also recognise the contractors' association as a useful body in offering advice on suitable contractors by their accreditation scheme, and there is also the possibility that the contractors' association could form a channel for support to emerging contractors.

A fair deal for road workers

While many contractors recognise that it makes good business sense to offer a fair employment 'package' to their workforce, there will always be some who will try to exploit them. The lack of work in an area resulting in large numbers of people desperate for work, increases the potential for worker exploitation. To counter this, many countries have labour laws which seek to guarantee certain rights such as a minimum wage. These labour laws may be flouted either intentionally or unintentionally by small scale contractors, while small firms, due to their diversity, method of operation and number of casual employees, may be neglected by labour inspectors.

When a government or controlling body sets minimum wage rates they must be realistic in relation to current market rates, and there should also be a mechanism to adjust the minimum wage level to take account of inflation as rates can quickly get out of line in high inflation economies. Setting a realistic minimum wage is a difficult process. If it is set too low workers can be exploited. On the other hand, a high minimum wage can mitigate against the choice of labour-based technologies by pricing them out of the market.

Clearly the primary reason for setting a minimum wage is to improve the living conditions of the workforce. However, these legislative measures are only effective if they can be effectively policed. Often it is better to promote a free market where wage rates reflect the genuine cost of labour, rather than setting a wage above the minimum amount people are willing to work which will discourage job creation and encourage the substation of capital for labour.

A fair deal for the community

The whole community are the ultimate beneficiaries from a new or improved road network, which facilitates links to services such as health care and education and also facilitates trade within the community and externally. Most communities would like improved or new road networks around their home area, so strict ranking criteria are required to determine which areas or routes should be addressed first. The criteria chosen depend on local conditions, but may include:

- the number of additional people brought within 500m, 1 or 2km of a road;
- the quality of the existing road network;
- the current level of traffic on existing road network in that area;
- the density of the existing road network (km of road per square km);
- the interest and support shown by leaders/councillors in the area for improved road networks;
- willingness to contribute fully or partially to the cost of providing and maintaining the facility; and
- the current journey time to the nearest market.

It is likely that a set of criteria for prioritising work will be required. Regardless of the criteria chosen, the community should be made aware of how the priorities have been chosen in order to maintain confidence in the road development programme.

Tools and equipment

Bank interest rates in developing countries are usually high, often reflecting a high level of inflation and currency depreciation (interest rates of 15 - 48 per cent were revealed in a recent survey).[7] The high cost of, and problems of securing, finance in emerging and developing countries and the utilisation-cost relationships mean that any equipment (whether heavy or intermediate) must be highly utilised to have a chance of paying back its investment and, for a contractor, to make a profit. The market for contract road works in most developing and emerging countries is particularly variable and precarious. Contractors must minimise the risk of having serviceable-but-idle plant.

There is therefore potential for the provision of intermediate equipment hire services, possibly through the road authority itself by utilising plant originally belonging to the authority's direct labour organisation. As the majority of larger items of construction equipment regardless of type (intermediate or heavy plant) must be imported and paid for in foreign currency but financed through a loan at interest rates of local banks, contractors may be tempted to purchase second hand equipment. As far as the government are concerned this approach can be a double edged sword. Suppliers can inflate the import price of second hand equipment in order to bypass foreign exchange controls or artificially raise their share in a joint venture. There is also the potential for equipment dealers to off load faulty or scrap machinery to unsuspecting contractors and road authorities alike. Governments can therefore ban the imports of second hand equipment or impose a punitive tax on imports. While this approach may eliminate problems described above it is also likely to restrict contractors access to legitimately imported equipment that is likely to meet the contractors needs at a fraction of the price of new equipment.

A role for local consultants
The 'traditional contracting' is a tripartite system consisting of 'Clients', 'Contractors' and 'Engineers'. The role of the engineer is to undertake feasibility studies, the design work and then supervise the construction of the project. Consultants are likely to be more efficient than government departments as they operate in a competitive environment. This environment encourages them to reduce not only their own costs but also the cost of the construction by effective use of local materials, willingness to adopt alternative technologies and changes in design procedures. Countries with an established private contracting capacity also have a competitive consultant sector. It is essential that the development of local consultants to undertake the 'Engineer' role are considered in parallel to the development of local contractors.

Small local engineering firms cannot be expected to manage large complex projects, where foreign engineers will be required. However, the majority of road construction, rehabilitation and maintenance projects are small and well within their capability.

Procurement and payment
Contractual relations should liberate the producer, but too often the contracts negotiated in the public sector are absurdly detailed. Instead of specifying the outputs, they describe in enormous detail exactly how the job should be done. The explanation appears to be that each party to the contract feels trapped by

the other. Purchasers, because they feel so insecure in their relationship with the providers of services, believe the only way to guarantee quality is to write in enormous detail exactly how the job should be done. If the producers of services genuinely felt that they had a variety of purchasers and if the purchasers knew that they could always buy from somewhere else this would give greater reality and credibility to the contract, which in turn would mean it did not have to be so over-precise.[9]

References

[1]ILO, (1993) *From want to work: job creation for the urban poor*. ILO, Geneva.

[2]Hibbs, John (1993) *On the move......: A market for mobility on the roads*. IEA, Hobart Paper No 121. The Institute of Economic Affairs, London, p 70.

[3]Lee, Douglas B. (1995) 'On the role of Government: Some recent revisions'. Paper prepared for the annual meeting of the Association of Collegiate Schools of Planning, Detroit.

[4]Hibbs, (1993) *op cit.*

[5]Carlsberg, Sir Bryan (1996) *Competition regulation the British way: Jaguar or dinosaur?* IEA Occasional Paper 97. The Institute of Economic Affairs, London, p 9.

[6]Miles, D.W.J. (1996) 'Towards guidelines for labour-based contracting: a framework document'. *MART Working Paper 1*. Institute of Development Engineering, Loughborough University, UK, p 29.

[7]Atkins, H. (1994) *Labour-based opportunities in construction*. Development Bank of Southern Africa Policy Working Paper 10 (Construction and development series Number 6), Halfway House, South Africa.

[8]Larcher, P.A. and Petts, R.C. (1997) 'Selective experience of training, contracting and use of intermediate equipment for labour based roadworks' *MART Working Paper 2*, Institute of Development Engineering, Loughborough University, UK, p 21.

[9]Willetts, D. (1997) 'Civic conservatism' p 133 in J. Gray and D. Willetts *Is Conservatism Dead?* Profile Books, London.

Chapter 3

An Enabling Environment

Markets work when there is an equitable business environment which enables purchasers and providers to trade to their mutual advantage. Where small indigenous contractors are disadvantaged, it is necessary to take steps to create an enabling environment which favours those companies which make a serious commitment to the industry. An effective enabling environment has many facets, including a fair fiscal regime, a reliable administrative and management framework and proper procurement procedures. In order to bring such an environment into being, it may well be necessary to draw on some measure of international technical assistance.

3.1 A positive fiscal regime

Lack of financial resources is the most common problem cited by small scale contractors as restricting the development of their business. While it is not sensible for governments to provide unconditional cash 'handouts' to contractors, they can ease the financial restraints on their business by addressing unhelpful economic and fiscal policies. For example, adverse currency exchange rates or excessive import duties can create severe problems for small contractors who have to obtain equipment or materials using foreign currency. Rapid changes in exchange rates can be particularly damaging. For example if a contractor took a loan of US $1000 when the national currency was valued at 500 per $1, the value of the loan in national currency units would be 500,000. If the following month the exchange rate was adjusted to the market rate of 1200 national currency units per $1, the outstanding loan would escalate to 1,200,000 national currency units (assuming that no repayments have been made).

Import duties

Fiscal charges for which businesses are liable include taxes, duties, levies and tariffs. Throughout the following discussion the term 'tax' will be used to cover any charge which contractor may face. When a contractor is looking to import equipment from overseas, import duty may well be chargeable and this will be an additional burden besides the foreign exchange risk. This may be as much as 50% of the purchase price of the equipment, but sometimes varies capriciously. For example in Ghana (at the time of writing) contractors can import agricultural trailers for hauling gravel and incur no import duty, because the equipment is predominantly for agricultural use, and the government has made a positive commitment to increasing the agricultural capacity within the country. However if a Ghanaian contractor imports a heavy duty construction/ gravel trailer or other item classified as construction equipment, a 15 per cent import tax is payable on top of the purchase price. These problems may be compounded in some countries by the delays that are experienced in the payment receipt and clearance required before the equipment can be removed form the point of importation. Even if the contractor pays the import tax immediately that equipment arrives in the country, it can take many weeks before the necessary paperwork is completed and a payment certificate issued to the contractor, so that the equipment to be released.

In terms of contractor development there is a case for removing or substantially decreasing the import taxes payable on construction equipment (although there is also a case for taxing capital-intensive equipment items in order to favour the use of labour-intensive techniques). Local contractors are definitely disadvantaged in those cases where they seek to compete with large contractors or foreign contractors on foreign-financed projects where equipment import duties are waived.

Visibility for taxation

Some governments have attempted to develop their construction sector by adopting policies to offer subsidised credit, preferential interest rates and tax incentives. However, these measures are usually limited to formal sector businesses which are registered for taxes and are subject to the full range of construction regulations. They do not reach those small contractors who operate in the informal or 'invisible' sector. Where they are large enough to be 'visible' and considered eligible for these policy benefits, the level of bureaucratic procedures which are necessary to qualify will often outweigh the benefits which can be derived from pursuing and obtaining the necessary paperwork and authorisation.

In addition to slipping through the net in terms of registered contracting companies, small contractors either legally, illegally or by default often slip through the tax net. They may not pay any form of tax to the government in respect of the work which they have undertaken. While this may be considered advantageous in the short term, problems are encountered as the business expands - the medium scale contractor growth trap. When a contracting business reaches a certain size, it becomes 'visible' to the government and is liable to pay tax on all its profits (including those that have been evaded in the past). This is likely to result in a financial crisis, and a need to install formal accounting and budgeting procedures as well as for the owner to manage the business in a less personal way. Contractors may find that once they reach this size, they are unable to expand or make sufficient profits to continue to grow. The more shrewd among the smaller firms may foresee these problems, and opt to stay small and thus remain invisible.

Clear fiscal policies

Governments should formulate clear fiscal policies which foster the development of new contractors and the growth of existing contractors. There must be clear policies on company taxation to make it appropriate for all sizes of company. As the administration of taxing small companies is likely to be very time consuming and result in minimal revenue, a policy of not taxing small companies with a turnover less than a particular threshold would be a positive step in supporting new and small scale contractors. It would not result in any significant loss of revenue compared to the current situation, but would mean that small contractors could operate more openly. When contractors' turnover reached the threshold level they would become liable to tax on a sliding scale. This sliding scale would enable the tax level to be progressively increase from zero to the country's standard level as the turnover and profits of a company increase from the threshold level. The tax threshold and sliding scale would have to be determined bearing in mind the economic standing of the country concerned.

3.2 Competent clients

A prerequisite for an effective enabling environment for a market in roads is to ensure that clients are commercially competent. Competence in this instance can be defined simply; the client must have reliable plans for work to be done and also have access to sufficient funds to commission contractors and consultants to execute it. The issues of how to raise finance to undertake road construction, rehabilitation and maintenance and to develop a contracting capability is discussed in chapters 5 and 6. This section will address ways in which client organisations can mobilise and manage their finances so as to achieve the greatest output from national resources.

Earmarking of specific taxes

There are two ways for a government to secure and manage the funding required to undertake their road projects; either earmarking of taxes from particular income flows or through a dedicated and independently-managed road fund. Funds are obtained by earmarking when various proportions of taxes or other revenues are set aside and assigned to road projects. For example, a government could decide that 25% of taxes on the sale of new vehicles and 30% of fuel tax levies should be allocated directly to a specific budget head for road construction and maintenance. Earmarked funds are obtained as part of the governments' overall tax scheme and are held in the main treasury account. The advantages of earmarking are:

1. that such funds tend to be more buoyant in that they rise automatically with inflation and economic growth; and

2. that it is less likely that funds will be put to other uses, since the diversion of funds would be more visible than a simple movement of funds between budgets.

However, as with earmarked cattle which are in danger of rustling, even earmarked funds are not completely safe from plunder by other departments who have already overspent their budget and which may be higher in the political pecking order than the ministry responsible for roads.

A road fund

A safer way of securing and managing the necessary finance for highway construction and maintenance is through the establishment of a dedicated road fund. The main difference between earmarking specific taxes and a road fund is that the fund is fed by tariffs, which provide a separately identifiable income stream from the general tax system, and the money is held in a separate account controlled by the road agency. The road fund should be managed by a board which is independent of government, and it should have the power to set the level of tariffs so that it can operate commercially by controlling or influencing all three business variables (price, volume and costs) rather than the usual dilemma of lacking any pricing option, but facing continuous rises in volume with income which is inadequate to cover a realistic level of costs.

A road fund of this kind provides the road agency with a more reliable financial source, which allows it to budget in the confidence that funds will be available to pay' contractors and consultants in accordance with contractual commitments. The agency is thus more accountable both to the government and to road users. The road fund should be set at a level which allows the road network to be realistically maintained and expanded. In principle all vehicle owners should contribute payments to the road fund, and these payments should be directly related to their share of the costs of providing and maintaining the highway network. In other words the charges should be related to costs resulting from road usage. In practice this will be difficult to achieve with any degree of precision, but it is desirable to establish a clear charging structure which can be understood and accepted as broadly 'fair' by all road users.

Regardless of the system used to obtain road maintenance funding, there must be an effective means of controlling the money collected to ensure that is it used as effectively as possible to maintain the road network. At the most basic level it must ensure that the revenue collected from road user charges is actually spent on the road network, rather than being absorbed into the general government budget. Money collected for road management would normally be put in a specially designated account which is referred to as a Road Fund. The Road Fund should be administered by a Road Board representing all stakeholders in the road sector. It must be independent from the government, although government departments may be represented, to ensure the actual needs of road users are addressed.

An autonomous roads authority

The management of the road network has traditionally fallen under the jurisdiction of the Ministry of Works, where it is seen as one of many areas of infrastructure. As a result it has been difficult to monitor the performance of the road network management in relation to the priorities of road users, since information and general performance data is 'lost' in the centralised running and accounting of the Ministry. By establishing an autonomous road authority the government is able to create an organisation which is run by representatives of road users and can be monitored closely against agreed indicators. With a clear command structure and specified objectives, the board can demand sound business practices which offer value for money. If this road authority is financed through a road fund then it is likely to be more financially accountable for its actions.

Major policy reforms rarely succeed without the support and commitment of key stakeholders. The World Bank Road Maintenance Initiative has undertaken a broad range of work in a number of countries in Sub-Saharan Africa, and has concluded that 'the key to policy reform is ownership' and that 'local stakeholders must agree on the need for reform, the reforms must be feasible, widely supported and should have more winners than losers'.[1] Heggie notes the importance of winning the support of the ministries of planning and finance and 'of utmost importance' the private sector, since the latter are the main users of the network and also pay for it. The composition of the road board gives an opportunity to involve as many as possible of the key stakeholders in the management of roads, 'which is usually a precondition for getting road users to agree to the introduction of a road tariff, since they are generally unwilling to pay for roads unless they can influence the level of the road tariff and the setting of spending priorities'.[2]

The roads board

Thus the road fund should be administered by a board on which all the major stakeholders in the road sector are represented. It should be independent of the formal government bureaucracy and its constitution should encourage it to focus on the felt needs of road users, rather than become a battleground where individual members each fight for their own functional or sectional interests. It is likely that the road board will include representatives from most of the stakeholders listed in Table 3.1.

Each of these groups will have their own agenda for spending the financial resources at the disposal of the fund, but the board collectively will have to recognise the need to compromise and work towards a generally acceptable solution that will optimally represent the interests of all stakeholders. Thus it is essential that the person appointed to chair the board should have the experience and confidence to ensure that board members pursue a coherent overall policy and work together productively to provide clear policy guid-

Table 3.1. Road board stakeholders		
Group	*Interest*	*Role*
Public transport operators Road hauliers Cyclists Pedestrians	Bus, minibus and coach passengers Long distance heavy load requirements Non-motorised transport Personal movement	Represent interest of specialist user groups
Contracting organisations Motor manufacturers and traders	Road maintenance and construction Vehicle requirements	Represent suppliers interest
Ministry of Finance Ministry of Planning Ministry of Works	Fiscal implications Long term economic planning Overall planning and co-ordination	Protect national interest
Chambers of commerce Environmental groups	Traders and manufacturers Environmental interests	Special interest groups

Table 3.2. Typical contributions by road board members	
Activity	*Typical contributions*
Communications	Internally between management and staff, and between staff. Externally with stakeholders. Information management and media relations.
Management and technology	Data collection and analysis. Appropriate technology and management systems. Contract procedures and standards.
Asset management	Valuation of road network. Strategic management to ensure optimal use of available funds. Setting appropriate tariffs.
Safety	Monitoring of accident data. Establishing operational safety procedures.
Human resources	Terms and conditions of service for agency staff. Payment systems for contractors and their staff.

ance to executives of the agency. The type of contributions that can reasonably be expected of road agency board members are indicated in table 3.2.

Inevitably some of the special interest groups will have conflicting interests and priorities for spending the financial resources at the disposal of the fund, but the board collectively should be in a position to represent the interests of all stakeholders. Although the Board should be constituted in such a way as to ensure a broad representation, it must be accountable to the general public through regular auditing of the fund account to ensure that the money has been spent effectively and none has been misdirected.

There will be many demands on the road fund's resources, but the primary function must be to finance the routine and periodic maintenance programme. The road board may decide to provide funds to meet the costs of road rehabilitation or construction, but this should only be carried out if there are surplus funds having paid for the routine maintenance budget. In order to obtain funds it will be necessary to impose the tax levies described above. It is likely that constitutionally these levies will have to be set by the Ministry of Finance. However the Road Board should advise the Ministry on the level of taxes to be imposed in order to ensure sufficient financial resources. The Ministry of Finance also has a position on the road board, which allows them to advise on the financial implications of the board's proposals.

3.3 Financial planning and control

The two most common problems experienced by contractors are a lack of work continuity and long delays in receiving payments, and both are at least partly caused by inadequate financial planning and management on the part of client organisations. Construction is generally recognised as a cyclical and unpredictable industry[3], but the cycle is often more extreme in developing than in industrialised countries. To some extent this reflects a less reliable or predictable income stream, but it is often possible to alleviate the extremities of the cycle by seeking to establish a steady level of planned construction and maintenance of the road network to give a 'floor' of regular work. This would improve work continuity and provide a sound financial base for planning to ensure money is available to pay for work completed.

This floor level of work would provide a basis for establishing the minimum financial requirements. Other activities and projects could be added to the workload in a descending order of priority within the budget allocation for each year. If further funds were available, possibly due to a delay or cancellation of a particular project, other activities with a lower priority could be added within the annual budget allocation.

When planning work the total annual budget must always be taken into consideration. Often the tendency is to prepare large programmes due to political pressure, which far exceed the available budget and form an unrealistic 'wish list'. It must be accepted that it is not possible to undertake all the work that is required immediately, and a list which gives projects in order of priority must be drawn up. Coupled with a forecast of the annual finance available, this list could also indicate the likely waiting time for a project's implementation. If this list was made available to contractors, possibly through local government offices, it would assist them to plan their own work and prepare bids for each item of work in their geographical area or speciality.

Timely payment

Apart from the lack of work continuity, delayed payments by government departments pose the largest problem to small scale contractors. The availability of timely payments to small contractors is the 'killer assumption' in the design of a contractor development project. As discussed above the main problem is often the lack of available finance. However, even if finance is available, payments may be delayed for a variety of reasons. Firstly contract supervision is often poor since officers designated as supervisors are based in head offices and visit sites rarely due to lack of transport. Besides the implications for quality, this results in the contractor having difficulties in obtaining and agreeing interim payment certificates. Once the payment certificate is agreed it usually has to be approved by a large number of individuals before it even can enter the bureaucratic system to the treasury for payment. This system exacerbates the potential for a certificate to be delayed or even lost.

We have suggested that the best solution is to turn the road authority into a road board with autonomous powers to operate as a quasi commercial enterprise. This will ultimately reduce the overall cost to the government as contractors will not need to use higher rates in order to cover their finance charges as a result of late payments. A simpler interim measure may be to pay interest charges on outstanding balances to contractors (as a counterpart to the liquidated damages that are claimed from contractors for late work). This would help contractors to mitigate potential losses and would help them to obtain an increase to the borrowing limits on their loan facility. It would also provide an incentive to government departments to improve their payment systems since the costs of delays would be directly exposed.

Decentralisation

An obvious solution to streamlining the acceptance of payment certificates would be to decentralise the payment system to regional offices, since the layer of bureaucracy at the head office usually provides only an illusion of useful control. Decentralisation offers two key advantages. Firstly, local offices have better access to work sites and government officials working in regional areas are more likely to be familiar with the projects which are being undertaken. Secondly, the long authorisation chain for certificates would be significantly reduced. Payment accounts could be set up with banks in regional areas to ensure that the necessary finances are available when certificates are submitted for payment. The alternatives for financing recurring road maintenance and ensuring timely payment to contractors are partial or complete decentralisation.

Partial decentralisation

The central road authority could determine the work to be carried out in each region on an annual basis. This could be based on information submitted by each regional road department. The agreed work could include finance for regular and periodic maintenance and road rehabilitation and construction. The management of the maintenance contracts would be carried out by the regional office and regular payments would be made into the decentralised regional accounts to cover these costs. Initially the regional offices could wait until the maintenance finance has been transferred to the regional account before they request tenders and commencing the maintenance work. However, once the system is operating smoothly and regular monthly payments are received from the central road account, maintenance contracts could be let more on a 'just in time' basis. Large projects can be controlled by the central office who would honour payment certificates authorised by regional offices. Regional offices would be able to monitor work progress on these large projects and request transfers to the regional road account when payment certificates were submitted. The central road authority should then immediately transfer funds to ensure payment to the contractors on time, and would receive a copy of the payment certificate to justify the money transfer.

Full decentralisation

In this case each regional road office applies for funds to carry out regular and periodic maintenance plus development of the road network, in fact all but the largest of road projects. The central road department assesses the 'bids' from each region and prioritises the work which needs to be done. They then divide the total annual road budget between each region according to the assessed priorities. Each region is then responsible for managing its own annual budget and receives a monthly payment from the central road fund. This option is likely to improve the flow of payments to contractors, but requires a high capacity in each regional office. It may be looked upon as a long term goal, which could be achieved gradually through partial decentralisation.

Effective audit

Regardless of the financing and payment system adopted there must be an efficient audit system to ensure that the money is used for the intended work. The systems described above promote an open and easily auditable system which will facilitate the monitoring of finances.

3.4 Operational management

Within many public bodies there is inadequate capacity to manage the procurement of road construction, rehabilitation or maintenance. This is largely due to an inappropriately structured and resourced bureaucracy, which lacks appropriate technical and managerial skills to meet the delivery requirements of the road users. There is therefore a need to promote a change process within these public sector organisations so that they are in a better position to manage the road network and other public infrastructure and amenities. In the development of a management framework to improve the capacity of public bodies to manage infrastructure, there are usually five types of constraint which must be addressed:

- inadequate productivity;
- centralised, top-down management;
- poorly paid and inadequately trained staff;
- management unrepresentative of the population (or end users); and
- a poor work ethic which leads to poor service delivery.

Road register

No business can be managed effectively without a current valuation of its portfolio of assets, and an estimate of the costs of maintaining them at an optimum level of repair in order to provide a satisfactory service to its customers. If a roads board is to be run in a businesslike way, the primary task of the management is to determine what roads form the network 'stock', their current value and the cost of bringing them up to an acceptable level. This data will form a road register and will include details shown in Table 3.3 below.

Complementary records should also be kept for the structures on the road network such as bridges and culverts. This data can then be used to assist the planning of maintenance programmes and determine future upgrading or new road scheme priorities. The register will also allow monitoring where there appears to be a design flaw in the road design, such as poor drainage provision, which results in a higher than average maintenance requirement. The management of a road register will assist in reducing the overall financial cost of managing the road network. It will allow maintenance to be targeted to the lengths of road where it is most needed, deferring maintenance on good sections of road. The register will also allow the planning of preventative

Table 3.3. Data to be included in a typical road register	
Group	*Interest*
Design and construction details for the discrete lengths of road	Type of surface Date of construction Thickness of pavement Organisation undertaking the work Cost
Maintenance carried out on the road	Specific details of work undertaken - eg spot improvements, regravelling. Cost of repairs
Records of road inspections	Defects and maintenance requirements Urgency of repairs Past records to enable a check on the rate of deterioration of the road

maintenance to be carried out before excessive deterioration necessitates more costly reconstruction.

The road register should be updated regularly to reflect the improvements to or deterioration of the network. It is likely that most inspection work will be required after the rainy season. The data could either be held on a computer database or on a series of record cards. The former approach enables the ready retrieval of key data such as:

- a list of roads which have not received any maintenance for over 3 years; or
- a list of roads which are under 5m wide.

However, a computer database is time consuming and costly to install. More seriously, if it fails to work effectively management will completely lose control of its information and will be unable to make rational decisions on priorities and resource allocation. A shorter term solution will be to maintain a record card system that will be quicker to set up and allow easy access to the data for all road authority personnel.

NGOs and communities

During the past decade their has been a rapid increase in the number of local non governmental organisations (NGOs) which are engaged in various forms of infrastructure provision. Many of these are supported by international NGOs and donor agencies, as they have the facility to operate more flexibly and effectively than many units within the formal government structure. These NGOs have primarily been involved in the provision of water infrastructure, but there is scope for them to become more actively involved in the provision and maintenance of the village road network in rural areas.

There are a number of advantages in utilising NGOs in road management. The primary reason is the contact which NGOs have with the local community. They are generally successful in reaching the poorest groups in the community and promoting community participation. Their comparatively small size and flexible management structure enables them to eliminate delays and inefficiencies caused by bureaucratic procedures, and allows them to adopt a flexible working approach and manage small scale work at competitive rates. Unfortunately the rapid growth in the number of local NGOs has in some cases resulted in weak organisational structures and insufficient technical and managerial expertise. It can also be difficult for the road agency to monitor the activities of a large number of unconnected NGOs and ensure their financial accountability. Programmes to utilise NGOs to assist with the management of the road network at the village level should be designed to ensure sufficient long term funding and support to promote institution building. Nevertheless the potential advantages of NGOs as front line executing agencies should not be overlooked.

Traditionally NGOs have been reluctant to work within formal government structures, and have seen their non governmental status as a positive advantage in gaining the trust of partners and beneficiaries. If an efficient management framework is to be achieved, the two groups must work together in a way that ensures that the autonomy of the NGO is maintained. The skills and strengths of each organisation must be utilised to the best advantage; including the NGOs' decentralised structures and access to local communities and the government road agency's intention to achieve compatible and effective services for the benefit of all road users.

Within the village or rural road network there may be a role for the communities themselves to play in the provision of the road network. Derbyshire and Vickers suggest that there are three factors which need to be present for community-based management of infrastructure to be successful:[4]

- a feeling within the community that there is a strong need for the services;

- a clear willingness within the community to take long term responsibility for operating and managing the services themselves (including a willingness to pay for the infrastructure where appropriate); and

- evidence of existing community organisation which includes leadership, motivation and skills.

Community contracts

Community contracts may provide a useful route to maintenance in areas where there is a lack of local private contractors and hence a market which is not fully competitive. The main perceived benefits of a community contract are in fostering local motivation to encourage sustainability, in that the community is directly affected by development of the road and therefore has a strong incentive to ensure the work is undertaken properly. Individuals within the community will also develop their vocational skills in road works, which will facilitate future road maintenance when it is required. The disadvantages of community contracts are that technical and managerial skills may not be adequate, so authorities may have reasonable doubts about the ability of communities to produce work of an acceptable quality. If a community

**Sustaining community based operation and maintenance –
Implications for practice**

- Participatory methods of working are vital. This means more than simply participatory contributions of labour and money in response to agency instruction, and implies continual processes of shared decision making between the agency and primary stakeholders at all stages of the project cycle, with the aim of developing the community members' sense of responsibility for and control over the local operation, maintenance and management system.

- Increase as far as possible the room for manoeuvre that community members have in relation to aspects of the service design - choice and location of service, and operation, maintenance and management system.

- Where community management is required, including capacity building as a project output, focus on developing skills in management, planning, analysis, decision making and problem solving . The time scales for construction work and capacity building are different. Capacity building requires separate resources of time, resources and personnel.

- Build in the issue of transfer of responsibility from the beginning of the project, with clear recognition that this is a process not an event.

- Recognise that tackling the problem of non-payment of recurrent costs is not simply a matter of adjusting payment levels, but should address all aspects of effective community management; institutional and technical.

- As far as possible, make change pay - i.e. creating paid jobs in service operation and management wherever possible and reducing or eliminating reliance on volunteer labour.

- Keep technology very simple to maintain and repair, with a reliable supply of spare parts and technical assistance available locally

Source: Derbyshire and Vickers, 1997

contract is being considered, it is essential that meetings are held with the community to agree the quality, cost and time frame for the work. The road department may have to provide direct technical support to the community or to commission a local NGO to assist the management of the project with the community.

Competent and motivated staff

Human resource constraints are a critical factor in the management of a road authority. Common constraints can be divided into four main areas:

- low morale due to very limited financial and technical resources;
- shortage of skilled staff, in terms of technical and managerial competence;
- low status, uncompetitive salaries and a poor career structure; and
- requirements to work under complex bureaucratic procedures.

The traditional career or promotion path for government officials in many countries is strongly biased to rewarding seniority and avoidance of blame for 'wrong decisions', rather than rewarding achievements and initiative. Staff are accordingly reluctant to suggest changes and propose new ideas, as they fear that their promotion prospects would be affected. An autonomous road authority would be expected to operate in a more businesslike way, and to base promotions and financial incentives on measured performance rather than length of service.

A shortage of skilled staff is usually addressed by providing training. However, the success of this approach depends not only on the quality of the training, but on the participants' working environment, their motivation and the relevance of the training to their present responsibilities and likely future careers. If trainees are unable to make use of the knowledge and skills that they have gained, they will become frustrated and the training investment will be wasted or even prove counterproductive. It is therefore necessary to develop a challenging environment for aspiring engineers, technicians and supervisory staff to ensure that they remain motivated and committed to the work that they are undertaking. A major motivation factor for staff, especially more junior ranks, is a reasonable remuneration package. Unfortunately it is currently a major disincentive, especially for engineers and managers. There has been a dramatic relative decline in the salaries of professional engineers in the public sector in many African countries over the past 25 years (see following box).

Engineers salaries: An example from Tanzania

"In 1972 the salary of the top engineers was on a par with the Chief Medical Officer and the Attorney General. The status of engineers has systematically gone down to the extent that in 1995, the salary of top engineers was placed at the entry point of junior doctors"

Source: Mwamila, University of Dar es Salaam 1997.

Lacking both status and rewards, it is not surprising that engineers sometimes supplement their meagre salaries through various authorised or unauthorised means ('moonlighting' or even 'daylighting', when they undertake private work during their formal working day, or - more serious still - various forms of corruption). The end result is that the "availability" of engineers to carry out their official duties is severely reduced. This is not only wasteful, but is also a constraint on overall construction industry performance, which has a particular impact on small emerging construction businesses.

Competent specialist consultants

Few developing countries have a strong local consulting sector, with the variety of skills that are needed to support a viable road contracting initiative. This is partly due to a previous emphasis on public sector employment, with young engineers expected (or even required) to seek employment in the public sector for a number of years after gaining their professional qualifications. Construction industry development projects have rarely targeted the development of local private specialist consultants, and it is difficult for local firms to emerge without some initial support and stimulation. Thus there is a case for initiatives to bring together individuals with strong qualifications, a range of relevant experience, adequate business skills and access to the funds necessary to convince clients that they are dealing with a reliable consultancy enterprise.

Internal co-ordination and collaboration

Although the road authority will probably remain within the public sector, there will be a need to change attitudes from traditional state owned enterprise characteristics to the more businesslike approach that obtains in the private sector. For example, most public enterprise managers think in terms of production and quantity; managers of private companies talk about quality,

Table 3.4. Characteristics of state-owned and private sector enterprises	
State owned enterprise characteristics	*Private sector characteristics*
■ Take time to ensures jobs are allocated to the right people at the right time	■ The available human skills are matched to the organisation's mission and goals
■ Policies aim to balance economic and social objectives	■ Policies aim to develop a coherent culture and balance current and future need
■ Rigid management structure	■ Flexible management structure
■ Planning is a reactive exercise	■ Planning is fully integrated
■ Group unity is emphasised for motivation	■ Individualism is emphasised for motivation
■ Protocol, rank and status are important	■ Informality and competence are encouraged
■ Education is an investment in prestige	■ Education is an investment in personal development

marketing, profits and sales. Ineffective communication is an enormous constraint not only to enterprise restructuring, but also to the introduction of modern management techniques and human resources management practices.[5] The road contractors undertaking the road authority's contracts will be adopting the private sector characteristics and will not tolerate the delays and complex procedures associated with the state owned approach (see Table 3.4). This change can pose problems to staff as the practical changes required will need to occur at a faster rate than the gradual adaption of organisational culture that would otherwise take place.

3.5 Procurement procedures

National policy and practices

The construction industry is often seen as an economic indicator, since fluctuations in construction demand are strongly influenced by the rise and fall of the national economy. On the other hand governments may seek to use the construction industry as a regulator of the economy, by increasing investment during periods of general depression and reducing investment when the economy is overheating with excessive demand and price inflation. While there is an economic case for this approach, it depends on an accurate reading of the respective strengths of the industry and the economy as a whole.

Experience suggests that governments rarely get these judgements right, and are more likely to exacerbate the construction cycle which will have a negative long term effect on the companies and institutions within the industry.

In addition to the demand fluctuations over a number of years there is an annual fluctuation in the work available. This is often due to the road budget allocation running from the beginning of the financial year. In many cases, allocations which are not used by the end of the financial year are 'lost' and must be bid for in the next year. There is often a rush to complete work at the end of the financial year, which reduces the effectiveness of the road budget. It should be possible to tender for work at the end of the financial year to be undertaken immediately the new allocations are available. This would even out the flow of work throughout the financial year. The weather patterns in some countries may also result in an annual cycle of work fluctuation if it is not possible to carry out work in the wet season. In order to maximise the potential working time, whenever possible contracts could be tendered during the wet season to start at the end of the rains.

Delivery mechanisms

There are many procurement methods available to undertake the construction and maintenance of roads. Apart from force account, where the government is responsible for the whole process, there are six methods for undertaking construction work using contractors.[6]

1. Traditional contracts

In a traditional contract the contractor will bid for work using a measurement form of contract. In other words the contractor will price a bill of quantities with estimated quantities against each item, and payments will be based on actual measured work. The employer, usually the government, will have selected and prioritised the projects which need to be carried out and is responsible for ensuring that sufficient finance is available to cover the project costs. Depending on the experience, workload and capacity of the employing organisation, design and supervision of the work can be undertaken in house or subcontracted to an external consultant. The contractor is responsible for implementation of the project, including materials management, construction management and the actual construction. Traditional contracts effectively foster an adversarial relationship between client and contractor, and therefore are not generally conducive to contractor development unless some form of contractor development agency operates in a third party role with links to client and contractor.[7]

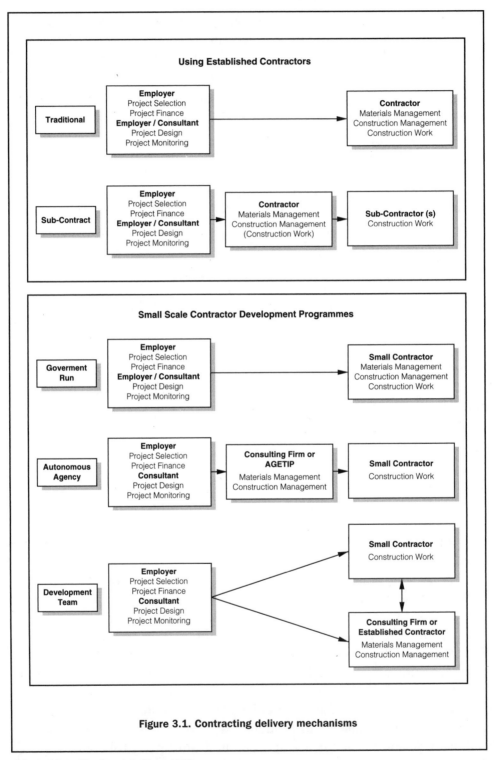

Figure 3.1. Contracting delivery mechanisms

Adapted from Stock and de Veen 1996

2. Sub-contracts

This system is similar in principle to the traditional contract, except that the main contractor will employ another firm to undertake some of the construction work. The scope for sub-contracting is usually defined in the main contract, and most contracts do not permit a contractor to sub-contract the whole of the work (although it may happen if the client's representative is not vigilant). Sub-contracting provides an avenue for new firms to establish themselves in the industry, since their legal and financial obligations are less onerous, and they can focus on their technical responsibilities. In return for these advantages, sub-contractors have to accept lower rates than those in the main contract (to provide a margin for the main contractor) and also run the risk that unscrupulous main contractors may not pay them in full for their services. These considerations may encourage clients to formalise the sub-contracting relationship by setting up a management contract, in which the management contractor may be required to adopt an explicitly developmental role.

3. Management contracts

In large projects a number of separate sub-contractors can be employed by the main contractor to undertake different parts of the project. This is a development of the simple sub-contract and may be referred to as a management contract. There is clear scope in this delivery mechanism for new small contractors to be employed as sub-contractors to the larger main contractor. In a management contract, the main contractor is paid specifically for management services rather than relying on the margin between the main contract unit rates and those negotiated with the sub-contractor. The management contractor is effectively in a professional consultancy role, and will be responsible to the client for negotiating a competitive rate with sub-contractors but may also provide training and advisory services to the less experienced sub-contractors in order to develop their skills.

4. Government-run contracts

This is a contractor development model based on the traditional delivery mechanism. In this programme the government agency is responsible for packaging works into suitable size contracts for small contractors, and may also provide training, advisory and other assistance to enable them to carry out their tasks effectively. The major difference between this system and the traditional contracting system is the need for the government to have an increased monitoring capacity, due to the number of contracts which will be let and the likely inexperience of the target group. The government agency will also be responsible for undertaking training and providing other support to the contractors. Thus supervisors will have to be encouraged to change from a purely 'policing' role to one in which they combine a concern for the continuing responsibilities for quality and time control with a more developmental and supportive approach. This change in approach may not in practice prove easy to achieve.

5. Autonomous agency

The major problem with the government-run contractor development system lies is the major reforms and capacity building that are required, and these can be eased by establishing an autonomous road agency. In this system the government hires an agency to undertake the contract management and contractor training aspects of the project. The government is still responsible for financing road projects and prioritising the work to be done, but pays the agency on a cost plus fee basis to manage the work including the payments to contractors. The perceived advantages of an autonomous agency are that the staff can be selected and paid a realistic salary to ensure well-motivated and high calibre organisation. Donor agencies are also better able to monitor and account for the money spent through a private agency. There are two possible disadvantages of utilising this system. Firstly, the private agency will be run to make a profit careful monitoring will be necessary to ensure that it does not abuse its monopoly situation. Secondly, this system does not promote a change in the government road agency itself, which may be an inherent weakness in the sustainability of this system or development approach.

6. Development team

This system is similar to the management contracting approach where a large contractor manages the materials and construction management aspects of a project. The small contractor will be responsible for the construction of the work. The major difference is that the small contractor is paid by the road agency rather than the development team, which will be paid under a separate agreement with the road agency. The contract with the development team requires the larger contractor/consultant to undertake mentorship with the small contractors to assist them to gain experience. Successive projects undertaken by the small contractors will result in them taking on more management responsibilities.

Contractor development

Each of the delivery mechanisms has a different approach to developing a contractor to undertake all the roles required to manage a business successfully. One extreme option is to make the contractor responsible for all aspects of the project (planning and materials management and construction of the actual works), but to offer small simple contracts in order for experience to be gained without a high risk. The alternative is for a contractor to be responsible for only the physical construction work, and to be slowly introduced to the issues of management once these vocational tasks have been mastered.

Contractor classification

While governments and Ministry of Works may be pressed to open work to contractors by allowing them to tender indiscriminately, the overall effect on the small contractors is often negative due to their limited financial, material and technical resources. The government or procurement organisation can categorise the contractors through a formal registration system into a series of classification levels according to their ability and resources (see following table 3.5).

In this example the firms enter the system as Grade 1 contractors and progress up the levels until they reach Grade 5, which represents a contractor able to undertake projects with the country's standard contract documentation and competing on the 'open market'. The contractors are not allowed to bid for work in a higher grade than one to which they belong, and may be prevented from bidding for work in lower categories (downward plundering). The principle of a classification system is that it protects both the client and

Table 3.5. Typical contractor classification system

Class	Contractor's skills and experience	Contract details	Maximum contract value
1	Some ability to organise. Basic artisan skills	Written instruction setting out the charge for labour wages, together with the contractor's mark-up and profit	$3,000
2	Established artisan. Civil engineering ganger, charge hand, gang boss.	Cost of labour, materials and basic equipment, including contractor's mark-up and profit, covered by a letter of agreement or simple contract	$10,000
3	Practising small contractor	Minor works contract with minimal bond requirements	$50,000
4	Established small contractor	Contract with schedule of rates awarded through open tendering	$200,000
5	Established medium-scale contractor	Contract with bill of quantities awarded under open tendering and requiring performance guarantees	$700,000

contractor. Clients are protected from employing a contractor who lacks the resources or experience necessary to undertake the project. On the other hand small emerging contractors can be protected from fierce competition by large well-established firms which could undercut the small contractor or even accept a short term loss in order to maintain their workload and keep staff and equipment employed.

The classification system allows progressively more testing levels of contract in which greater risks and responsibilities are placed on the contractor. In the example above the level of performance bonds required and contract value are increased as the contractor gains experience.

If a contractor classification scheme is adopted for contractor development, a clearly defined set of criteria must be established to determine:

- how a contractor should be registered;
- when a contractor can or should move on to a higher level; and
- when a contractor should be demoted or removed from the classification.

The overriding factor for success in contractor classification schemes is the requirement of an open and fair system. Registration and classification may be done by the Ministry of Works on behalf of most clients, or could be carried out by an independent organisation such as a National Construction Council or a Contractors' Association. The initial registration of new contractors would be relatively risk-free as they would automatically be required to commence at the lowest level. However, there should also be a mechanism for existing contractors who wished to join the scheme to enter at a higher level depending on their experience. In order for the scheme to be successful the contractor must see that there is an opportunity to rise up the Grading in order to be able to undertake larger and potentially more profitable jobs. A rise in the Grade should not just be a time serving exercise, but should be based on performance and capacity, taking into account defined criteria such as:

- financial capacity (working capital, turnover and value of assets);
- number of personnel including experience and qualifications;
- number of years the company has been operating successfully;
- the turnover of the company over successive years; and
- the value of the tools, equipment and other assets owned.

Care must be taken in establishing these criteria. For example, excessive emphasis on the ownership of fixed assets may encourage contractors to purchase unnecessary plant when more suitable equipment is available for hire. In addition, if the government's contractor development policy framework favours the use of labour-based contractors to an excessive extent, promising firms could be effectively trapped in the lower categories and unable to rise to a higher grade. Thus criteria could be modified to measure a mix of more relevant indicators such as the value of equipment owned, equipment hire charges incurred during a year or number of labourers on payroll.

Many small contractors have an erratic workload with long periods between contracts. They are therefore effectively 'occasional contractors'. The organisation in charge of administering the scheme should therefore be in a position to assess whether a contractor is no longer operating in the construction business or is simply 'between contracts', but still interested in the construction business and capable of remobilising quickly should work become available.

Lessons from experience

There are a few pitfalls which have been encountered regularly when contractor classification schemes have been implemented. In order to be successful all contractors should register in the classification scheme to prevent inexperienced and under-resourced contractors from unfairly competing for work. Registration is most likely to be accepted by contractors if the scheme is promoted or adopted by the client organisation and contracts will only be offered to registered contractors. Clients will have the added security that they are offering work to contractors who are capable of completing their work.

During periods of low work availability, especially for higher value contracts, some large contractors have bid for work offered to lower grade contractors. It is therefore advisable to set lower limits, in addition to upper limits, that a contractor can bid to prevent "downward plundering" by larger firms. For example a grade 3 contractor could be restricted to bidding for work in grade 3 or grade 2, which would protect new contractors in grade 1.

Finally if classification criteria are based on monetary values, they must be adjusted frequently enough to take into account the inflation levels in the country.

Contract packaging

If contractors are graded into bands there must be sufficient work available within each grade to offer a sufficient market. Some large contractors may offer sub-contracts to small contractors to undertake building works, such as culvert and masonry channel construction or minor bridge repairs. It is unlikely that these ad-hoc contracts would provide sufficient work availability for contractors in the lower grades. Another mechanism is required which offers sufficient numbers of small and medium scale contracts.

Road construction and maintenance is a linear, repetitive construction process. With a few exceptions at major structures, the construction or maintenance work is the same along the length of the road. This allows road construction or maintenance contracts to be easily 'sliced and packaged' into smaller contracts. The example below shows how a 40 km road maintenance contract could be packaged into six smaller contracts suitable for the lower grade contractors.

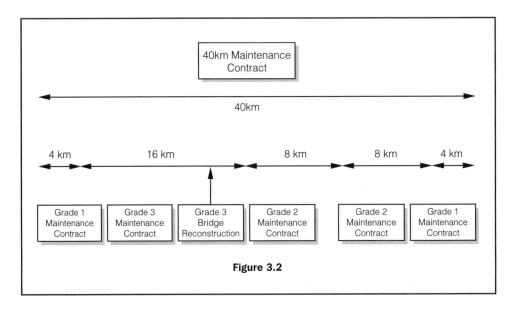

Figure 3.2

One common objection to breaking up large contracts into smaller ones is the increase in administrative and supervisory work. However, the workload will not increase sixfold as the contract documents would be broadly similar, requiring only minor modifications. The tenders for each contract could also be staggered by a week or two, to prevent an overburdening of the contract administration office. In general, the current level of site supervision on road contracts is poor due to the lack of resources and suitable personnel to monitor large projects. By splitting one contract into smaller packages the amount of time required to oversee the group of contracts is likely to increase. However, the contracts will be simpler, and could be supervised by more junior staff situated more locally and therefore make better use of the limited resources available. In addition to supervising work carried out and checking work for payment, these supervisors could also compile and maintain a road register to record the condition of the road network in order to plan and programme future maintenance.

An alternative to slicing and packaging the work into sections is to divide it into different types. Contracts could then be offered to contractors who are suitably qualified or experienced to undertake specific types of work. Using the example shown above the maintenance project could be divided into four types of contract:

Simple contracts: including activities such as grass cutting, culvert cleaning or de-silting and side, turnout and cut-off drain maintenance. To undertake this work a contractor would require little investment in equipment, as only a few simple hand tools and a supervision vehicle would be required coupled with a working knowledge of drainage and vegetation maintenance. The contract would not require bonds or sureties, and would be paid on a simple visual inspection to ensure that the work had been completed. In order to be successful the contractor would need to hire and manage the workforce, ensure timely payment of wages and organise suitable hand tools. This contract is essentially a labour-only contract as the majority of the contract price would be spent on labour wages.

Technical contracts: suitable for more experienced contractors who had a some simple equipment, such as a compacting roller, gravel trailer and tractor. The work would include pothole patching, resealing of pavements, regravelling small sections of road, small culvert reconstruction and possibly light regrading. The contractor would need to be more experienced in labour and equipment management compared to a contractor undertaking simple contracts. In order to enable contractors to move from undertaking simple contracts to technical contracts, tender regulations should permit the use of hired or borrowed equipment.

Gravelling and reconstruction contracts: involving the regrading, reconstruction of road surface and regravelling extensive lengths of the road. The contracts would be let to contractors with experience of road maintenance and access to the equipment required to carry out regravelling work (either owned or hired).

Speciality contracts: including bridge, culvert and other structural maintenance and reconstruction. The knowledge, experience and equipment required is fairly specialised, which precludes the inclusion of this work in the other contracts. Contractors working in this sector may have a range of experience and the contracts would reflect this experience. It is likely that contractors entering this sector of work would have already undertaken projects in other branches of civil engineering.

Local capacity building

Regardless of the delivery mechanism there will be a need to undertake a number of steps in order to establish and operate an effective and responsive contracting system. Depending on the delivery mechanism adopted different organisations would be responsible for the various steps. A typical series of activities is set out in the following box.

Twenty steps to contractor development

1. Appoint project manager.
2. Agree timetable for implementation of contracting process.
3. Identify suitable project 'packages'.
4. Develop contract administration team (see Chapter 4).
5. Determine delivery mechanism.
6. Develop contracting prequalification questionnaire.
7. Advertise scheme locally and nationally in press, radio and other suitable media.
8. Invite respondents to initial meeting, at which benefits and responsibilities will be explained.
9. Evaluate potential contractors.
10. Assemble list of prequalified contractors.
11. Provide training (if required).
12. Prepare tender documents.
13. Prepare contract documents including specifications.
14. Invite tenders from prequalified contractors - all prospective contractors should have the same information and time limits for submission of tenders.
15. Receive tenders and confirm receipt.
16. Review submitted tenders.
17. Award contract.
18. Supervise and monitor work to ensure quality, and certify monthly payments.
19. Inspect finished work and issue completion certificates.
20. Review and evaluate project.

The scope for joint ventures

The overall objective of classifying or rating contractors and packaging contracts is to increase the local contracting capacity. Small, simple contracts allow new contractors to gain experience without being committed to a large investment or the risks involved in undertaking a large project. While contractors are able to gain technical experience through working on road maintenance projects, it is considerably harder and will take far longer to improve their financial and risk management and tendering expertise. This learning process can sometimes be accelerated if inexperienced contractors operate in joint ventures with, or as sub-contractors to, more established, larger contractors (although there is an obvious danger of exploitation of the smaller firms by the larger).

In principle a contractor ought to learn more through a joint venture than a sub-contract, where the contractor is essentially undertaking the same work but for a different client. Sub-contracting should therefore be looked upon as a mechanism for the smallest and new contractors to gain experience. Joint ventures offer the less experienced contractor the opportunity to be involved with the various aspects of construction management. Joint ventures may be initiated between a large and small local contractor or between a local and foreign contractor. There is often potential for a more successful outcome between two local contractors as they are more likely to understand the problems and constraints under which the other party is forced to work. In order for a successful joint venture to be achieved between a local and a foreign contractor, the problems of lack of trust between each organisation must be overcome. Foreign contractors can often view local contractors as a liability if they are doubtful of how they are likely to perform, which may reflect badly on themselves. Local contractors are often keen to lead a project despite their lack of contract management ability.

Decentralisation

The decentralisation of contract administration allows local staff to respond rapidly to problems which are encountered on site. Road authority staff 'on the ground' will have a clearer idea about the different work which each contractor is undertaking including the specific difficulties or peculiarities of each contract. They will also have built up a closer working relationships with the contractors and their staff, which will facilitate working together to settle disagreements before they become large disputes.

When determining the degree of decentralisation, care must be taken to ensure that a level is chosen where there is sufficient financial and management experience to undertake contract administration effectively. There may be strong opposition from government officials in central offices, despite the attraction of a reduced workload. This may be due to corruption as well as an unwillingness to change location. By splitting work into smaller contracts and devolving day to day administration from the central office, there is less scope for improper inducements.

3.6 Appropriate technical standards

Many countries still lack suitable local civil engineering codes of practice. This lack of local standards often results in the construction of roads to a standard far in excess of that required for the likely level of traffic. This situation results in two negative effects, the costly importation of excess materials and equipment and an inhibition on the competitiveness of small local contractors. One obvious example of this problem is the design speeds for rural roads, which dictates the minimum radius of the horizontal and vertical alignments of the road. The use of unmodified standards from industrialised countries results in the selection of excessive design speeds, and hence large cuttings and embankments to enable the required alignments to be achieved. Such designs result in unnecessary movement of large quantities of material which require large items of plant; a situation which is biased against small local contractors and effectively rules out the use for labour based contractors.

National codes of practice

It is essential that national codes of practice are developed which take account of the materials and requirements of the country concerned. These national codes may be based on existing codes of practice from other countries, but must be adapted to take account of the prevailing conditions in the country concerned. The development and drafting of these codes should be carried out in association with all the stakeholders involved in road use, construction and maintenance including the road agency, road users, contractors, consultants, construction material suppliers and manufacturers and appropriate research organisations.

The use of foreign codes often precludes the use of local materials. As construction materials form a large part of the construction cost of a project, it is essential that the standards address the use of local materials for construction. This includes raw materials such as stone, aggregate and gravel, but also manufactured materials such as paving blocks, bricks and locally manufactured cement. For example it may be possible to relax the strength requirement for concrete if the building element is increased in size to compensate for this reduction in strength. The promotion of local materials will require research into the properties of these local materials and development of standard testing procedures. It also requires an investment to investigate the availability and location of raw materials and improved information on the quality and performance of manufactured materials. The research organisations involved in this materials investigation work should be included in the group responsible for the preparation of national codes and standards.

Availability of materials

In addition to drafting standards for the use of local materials, the availability of construction materials (both local and imported) should be addressed. Small contractors often claim that they cannot procure materials because either the government has a monopoly on the sale (or controls the supply) or that larger firms are favoured by materials suppliers. Government departments may even stockpile materials during shortages, which further restricts availability. Some governments have attempted to improve the supply situation by legislating to control the price of construction materials, but this is rarely effective and generates pointless bureaucracy. The best solution is to allow a free market for the supply of materials, encourage government departments not to stockpile materials and require suppliers to publish material prices in the local currency.

3.7 Appropriate tools and equipment

Tools and equipment can be defined as 'appropriate' if they allow road construction and maintenance to be undertaken in the most economic way, in other words at least cost. Relevant costs include labour wages, capital and running costs and possibly environmental or global costs. For example, if it costs a contractor $500 per day to hire and operate a mechanical excavator, how will the production costs vary if this $500 is spent hiring a labour force to undertake the same work instead? It may be argued that construction mangers will always choose the most appropriate method. However, this choice may be distorted by:

- incorrect information;
- errors in productivity calculations;
- equipment prejudices; or
- political pressure.

The share of wages in the cost of labour-based road construction and maintenance is generally in the range of 30-50 per cent.[8] The competitiveness of these methods is therefore heavily dependent on the productivity of the labour force. Although it is logical to suppose that tool quality affects the productivity of labour-based works, there was very little evidence to quantify the difference in productivity achieved by using tools of different quality. Even when agreed standards and specifications are available[9], poor quality hand tools are still being used in many road projects (and locally-manufactured tools are often of poor quality). Most of the tools used for road works are also used in the agricultural sector, but the harsher environment of road construction results in poor durability and a short life for tools. Few contractors or road agencies fully appreciate the long term economic benefits of investing in good quality hand tools.

The results of a survey undertaken within the MART initiative showed that the main criteria for procuring hand tools are relative availability and immediate cost.[10] Within sub-Saharan Africa, this favoured poorer quality tools imported from parts of Asia. If a choice of brands is available, good quality tools are only chosen if experienced technical staff manage the procurement process. However, non-technical procurement officers and tender boards tend to choose the lowest cost tools irrespective of quality. The survey also showed that new tools are rarely tested for conformity to specification, other than a brief visual inspection. The move from direct labour-executed road works to private sector contracting might have been expected to lead to a more rational choice, but there is evidence that road construction and maintenance projects which seek to promote the use of private contractors still frequently utilise poor quality tools. Interviews with contractors suggest two reasons for this:

- Many government-initiated contracting projects procure tools centrally on the criterion of least immediate cost, and then sell these on to contractors.
- Contractors frequently do not appreciate that significantly higher levels of productivity could be achieved with better tools.

Figure 3.3 below summarises the effect of continuing use of poor quality tools. The issues on the left of this diagram are well known, and a large amount of work has been carried out in an attempt to improve the initial quality of tools. The MART programme has also highlighted the neglected problems on the right of the diagram. In some areas of Africa there is little effective commercial demand for higher quality tools which are suitable for road works. The uneconomical distribution of these higher quality tools in small numbers through scattered retailers results in high unit prices, and encourages the purchase of poorer, but cheaper, tools. Productivity research

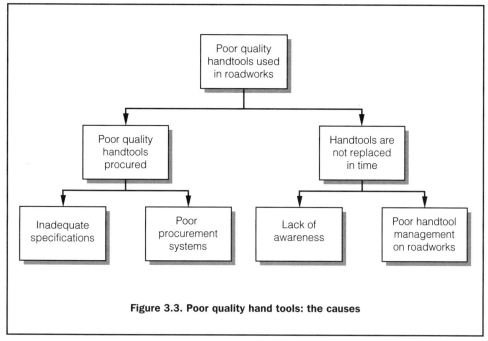

Figure 3.3. Poor quality hand tools: the causes

Source: Taylor 1997

undertaken by MART has shown that the immediate difference in output between new tools of different quality is very small. However, the productivity of the poor quality tool drops off quicker, and after a shorter period, than a better quality tool. For example the use of a poor quality, worn shovel for ditching work reduces long term productivity by 22 per cent. The research has also shown that few contractors are aware of the decline in productivity, and hence profit, when badly worn hand tools are employed.

Intermediate equipment

As labour-based techniques gained favour, it became clear that there was a need to support the labour force with basic equipment for certain tasks which are less suited to the labour-based approach. These include material haulage for distances greater than 200 metres, compaction of the road surface and maintenance grading of unsealed roads. The conventional solution might be to bring in heavy construction plant to excavate, move and regrade material. However, intermediate equipment, based on the agricultural tractor, can often prove both more effective and more economic. Furthermore it can often be locally manufactured, or at least locally repaired and maintained, and is sufficiently versatile to be put to other remunerative uses when not required for road construction or maintenance. The object of promoting intermediate equipment is not to displace labour but to complement it. An optimal mix of

Table 3.6. Two equipment approaches	
Heavy construction plant	**Intermediate equipment**
Sophisticated civil engineering equipment designed for, and manufactured in, high-wage, low-investment-charge economies. Expected to operate with close support and high annual utilisation. Usually designed for a single function with high efficiency operation.	Simple equipment designed for low initial and operating costs, durability and ease of maintenance and repair in the conditions typical of a limited-resource environment, rather than for high theoretical efficiency. It is preferable if the equipment can also be manufactured or fabricated locally.

Source: Petts, 1997

labour and intermediate equipment can provide a socially, economically and technically superior solution to a heavy equipment-based approach (see Table 3.6).[11]

Dangers of interdepenence

The definition of heavy construction plant highlights some problems which are commonly encountered when conventional plant is used in developing countries. Each item of specialised equipment has a dedicated function which results in an interdependence. If one item of equipment breaks down, then the whole fleet becomes idle as 'a link in the chain' has been broken. In addition the equipment often relies upon sophisticated mechanics, hydraulics and electronics to be fully operational. The specialist spare parts as well as the maintenance and repair skills for these items of equipment are only available, at best, in capital cities. All equipment and spares have to be imported which consumes scarce foreign exchange and, in the case of spare parts, may take weeks or even months to obtain. Plant manufacturers also produce model improvements and cease stocking spares for old and 'obsolete' equipment.

Second hand equipment

Contractors may choose to buy second-hand equipment which is available as a result of being imported for use on a larger project, the price being determined by the relative availability. Although the initial investment by the contractor will be smaller, this advantage may be offset by additional problems when utilising the equipment due to the need for additional spare parts, which may take weeks or even months to procure. The purchaser is also unlikely to have any information about how the equipment has been used, or abused, by the previous owner. The relatively small markets for construction machinery in each country result in few dealers able to provide the after sales support, specialist repair and maintenance skills required to operate this machinery.

Affordability

Heavy plant is particularly unaffordable for small contractors. They require equipment which is not only cheap to purchase, but also robust and easily maintainable in a modest rural workshop. By developing designs that could be manufactured locally, direct employment opportunities can be created and the lack of support services overcome. There are also economic arguments for the development of intermediate equipment. Many developing countries are subject to high interest rates, especially for the purchase of construction equipment which may not earn a reliable return. These high interest charges, coupled with the high capital cost of heavy plant, makes their use on all but the largest project economically unjustifiable. The argument is based on the utilisation requirement in terms of number of hours per year, which the equipment is required to operate in order to cover its fixed and variable costs. While the operating costs may be virtually constant regardless of the number of utilisation hours, the finance costs and depreciation increase dramatically as the number of utilisation hours decreases. The graphs below compare costs of using a motor grader and a tractor with a towed grader based on a 12-year economic life and an interest rate of 20 per cent.[12]

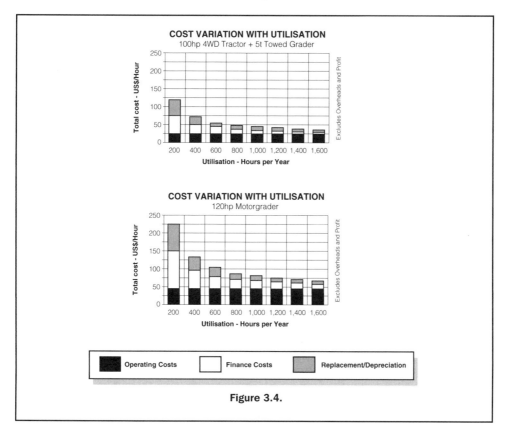

Figure 3.4.

Utilisation rates

A contractor operating in a typical industrialised country would expect to achieve a utilisation of about 1000 hours per year. At this point on both graphs the cost variation against utilisation is small due to the high proportion of the operating cost in the total price. However, a contractor in a developing economy would be lucky to achieve a utilisation of 400 hours due to cyclical demand and unpredictable delays. At this level of utilisation the cost of the motor grader is significantly higher than the tractor and towed grader. The total cost per hour also changes significantly for changes in the level of utilisation. These graphs highlight the problems faced by contractors working in low income countries when they attempt to utilise sophisticated plant.

Local supply

If a contractor makes an investment in an item of equipment, the investment will only pay off if maximum usage is obtained from that equipment. It is considerably harder to achieve high utilisation rates from specialist equipment than with more versatile equipment. The issue which must therefore be addressed is the local supply of the appropriate types of equipment. For example a contractor equipped with tractors with ancillaries such as compacting rollers, gravel trailers and towed graders, would possess a versatile complete set of road maintenance equipment at a fraction of the cost of a dedicated equipment fleet. When the contractor was unable to utilise this equipment, particularly tractors, on road contracts, it may be possible to hire them out in other sectors such as agriculture and irrigation work. The business would therefore be more likely to achieve the utilisation rates required to cover ownership costs.

Equipment hire services

As contractors must try to avoid having serviceable-but-idle plant, there is the potential for the provision of intermediate equipment hire services. This could be provided by dedicated plant hire firms, or by contractors hiring out their equipment when they have insufficient work themselves. This flexibility should increase the utilisation of individual items of plant, and therefore result in lower overall costs. Before commencing a plant hire operation, careful consideration must be given to the anticipated volume of demand for equipment. To be effective a plant hire company must be local to a contractor, however the demand may be too geographically spread out to make a plant hire company economically justifiable. If the hire company offered basic and versatile equipment, which widened the potential hire market, it is more likely that it would be financially viable. Initiatives will probably be required to convince road authorities, contractors and international agencies of the potential benefits of such an approach using intermediate equipment. Pilot schemes should allow the potential, technicalities, costs and benefits of such an approach to be established.

It may be possible to create a plant hire enterprise utilising the equipment from the existing public works department equipment fleet. This private company could hire serviceable but idle equipment not only to the public works department, but also to private contractors. The main issue to be addressed if this approach is adopted would be the level of control over the company between the public works department and private sector. The plant hire organisation should have sufficient independence to be able to operate commercially enabling it to:

- address the market demand;
- undertake rigorous management (maximise income, minimise costs); and
- implement a long term strategy and development plan.[13]

Criteria for choice

Heavy construction plant will still continue to be justifiable on many large, paved main road, reconstruction and rehabilitation projects where the factors of heavy traffic, sophisticated technical specifications, high guaranteed plant utilisation, economies of scale, intensive management, rapid implementation and relatively simple logistics favour a large-contractor, capital-intensive approach. However for most other road works in developing countries the use of intermediate equipment and labour is often cheaper and more appropriate.

3.8 Co-ordination of infrastructure investment

There has been a recent trend to develop sector investment plans in an attempt to improve ownership and co-ordination within government departments for donor support. This is particularly important where there are a number of different donor programmes running concurrently in the same country, since they all require significant resources from the government departments to monitor and staff these separate projects. Sector investment plans should therefore have a number of characteristics in order to improve the effectiveness of the donor support (see following box).

Characteristics of Sector Investment Plans

1. The operation should encompass all activities with significant expenditure relevant to the sector and the government should define the approach.

2. The plans require a coherent sector policy framework produced by the government providing at least a clear set of principles, particularly over the roles of the private and public sectors.

3. Local stakeholders must be in the driver's seat - this is very difficult to define and implement but starts with the government production of a concept paper for discussion by a new structure created to determine policies with representatives from all stakeholder groups.

4. All donors should sign on and the government itself must be committed to the process.

5. Common implementation arrangements must be agreed and accepted by donors for accounting, budgeting, procurement and progress reporting, preferably through strengthening local capacities.

6. There must be a major emphasis on capacity building with minimal long term foreign technical assistance.

Source: Harrold et al 1995

As yet, only a handful of sector investment plans are in place in Africa, and these are largely outside the field of rural infrastructure. However, initial experience has been reported to be encouraging.[14] The donor community has an important role to play in facilitating these reforms by only providing funding for road and other rural infrastructure in the context of a sustainable national policy.[15]

References

[1]Heggie, I.G. (1994) *Commercialising Africa's roads: transforming the role of the public sector.* World Bank, Washington, DC.

[2]Heggie, ibid.

[3]Hillebrandt, P.M. (1985) *Economic theory and the construction industry* (Second Edition). Macmillan, London.

[4]Derbyshire and Vickers (1997) 'The sustainable provision of poverty focused rural infrastructure in Africa: a study of best practice', DFID report, (unpublished).

[5]Prokopenko, J. (ed.) (1995) *Management for privatization: lessons from industry and public service.* Management Development Series No. 32. ILO, Geneva. p 284.

[6]Stock, E.A. and de Veen, J.J. (1996) *Reforms and delivery mechanisms for expanding labour-based methods in Africa,* SSATP Working Paper No. 347, World Bank, Washington.

[7]Edmonds, G.A. and Miles, D.W.J. (1984) *Foundations for change: aspects of the construction industry in developing countries,* Intermediate Technology Publications, London.

[8]Dennis, R. (1997) 'Results of a questionnaire on handtools for labour based roadworks'. *MART Working Paper 8.* Institute of Development Engineering, Loughborough University, UK.

[9]ILO (1981) *Guide to tools and equipment for labour-based road construction.* International Labour Office, Geneva.

[10]Taylor, G. (1997) 'Effects of poor handtools on worker productivity in labour based roadworks'. *MART Working Paper 9.* Institute of Development Engineering, Loughborough University, UK.

[11]Petts, R. (1997) 'Tractors in Roadworks' *MART Working Paper 7.* Institute of Development Engineering, Loughborough University, UK.

[12]Petts, R. (1997) ibid.

[13]Lantran, J.M. and Debussy, R. (1991) *Setting up a plant pool: Contracting out road maintenance activities. Volume III.* World Bank, Washington DC.

[14]Harrold, P. et al (1995) *The broad sector approach to investment lending: sector investment programmes,* World Bank Discussion Paper 302, World Bank, Washington, DC.

[15]Heggie, I.G. (1995) *Management and financing of roads: an agenda for reform,* World Bank Technical Paper No 275, Africa Technical Series, World Bank, Washington DC.

Chapter 4

Purchasers and Providers

The institution responsible for the road network would have to have a multi-disciplinary staff who would be capable of contract administration and supervision, project planning, financial control and policy making. In order to effectively perform these functions the road authority should have links to contractors' associations, the consulting profession, government ministries such as the ministry of works, road hauliers groups and chambers of commerce. This supports the argument in the previous chapter for a road board to oversee the activities of the road authority. The road board can also facilitate a flow of information and opinion to the road authority from the 'road consumers'.

4.1 Stakeholders and systems

In Chapter 3 we discussed the various measures that are required to foster an enabling environment to support a market for small scale contracting. This chapter seeks to assess the current and future availability of contractors and supervisors for both contracting and road authority groups, and then discusses the availability of a support framework to assist these organisations and measures to increase their capacity. It is logical that this chapter should follow the discussion of issues concerning the enabling environment. Programme designers often create projects which develop the purchasers and providers, but fail to address the fundamental issue of creating a market. It is essential that privatisation and contractor development is demand driven, so that there is a real need for contractors and a willingness to pay for their services. Programmes are doomed to long term failure if they create a private contracting sector (supply driven) but have not addressed the factors which create a market.

The primary stakeholders of the construction industry in developing countries are the same as those of developed countries, although the 'balance of power' between them is likely to be significantly different. The diagrams below symbolise the difference in strength of the stakeholders between developed and developing and emerging countries. The first point to note is that in developing countries the client is predominately the government, while in developed countries the client can often be from the private sector. The lack of resources and experience of contractors in developing countries places them in a much weaker position than the government and client. Under the traditional contracting system contractors are therefore forced to accept a proportionally greater contractual risk than they can reasonably bear. Finally, an indigenous consulting profession hardly exists in many developing and emerging countries (particularly in Africa).

An unstable system

In developing countries the tripartite system of client, consultant and contractor does not function in a stable way, as the consulting engineering profession

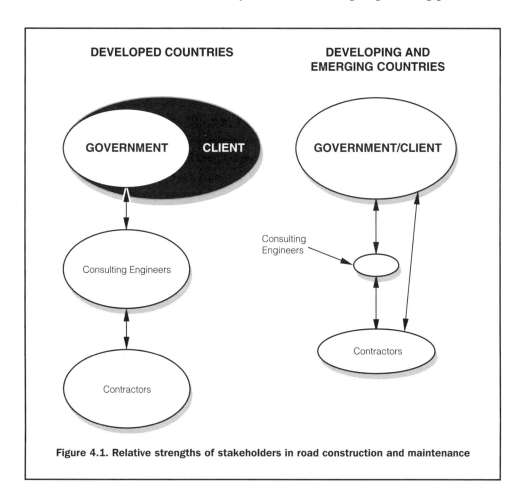

Figure 4.1. Relative strengths of stakeholders in road construction and maintenance

is often relatively weak. As a result, the client often takes on the main roles of the consulting engineer. This implies that government departments carry out the design work and then prepare, tender and supervise the contracts for work that they require to be undertaken. Within the existing state owned enterprise system there is in general the capacity within the road authority to undertake the design work. However, the outputs from the design group are then passed to the labour gangs to execute the work and to ensure that it is constructed to agreed standards. There is therefore little experience and capability to prepare and administer contract documents, supervise independent contractors and measure and certify work completed for monthly payments.

While the tripartite arrangement between client, contractor and consulting engineer forms the core of the institutional framework within developed countries, there are many other support organisations which are essential to enable the industry to function effectively. There are also other organisations which, while not essential, support the industry and enable it to develop new techniques and materials and improve productivity. Within developing countries these organisations rarely exist or are very weak with insufficient resources to provide a significant level of support. In developed countries many of the organisations in the support framework are financed (directly or indirectly) by the three stakeholders in the industry. Within developing countries this financial resource is not available as the engineering profession is often weak, contractors have very limited financial resources and government funds are insufficient to meet the requirements of the road maintenance budget, let alone support the industry framework which carries out the maintenance.

Two problems

There are therefore two distinct institutional problems which must be addressed by countries which aim to utilise the private sector to undertake road construction and maintenance:

- the lack of a consulting engineering profession; and
- the poor capacity within the support framework to assist the contracting sector.

In addressing the first issue, the long term objective should be to develop the engineering profession. However it is likely to take many years before the profession has sufficient capacity to undertake the roles which would be demanded from it in a developed country. As a design capacity exists in the government agencies the most appropriate solution in the short to medium term will be to develop the government agencies' capacity to prepare, award and administer contracts. Regarding the second issue, the contracting sector is generally poorly developed, partly due to the lack of a support framework for the industry. It will not be possible in the short to medium term to put in place an extensive support structure, but at least a lower level of support which addresses the core needs should be provided.

4.2 Clients

As the capacity of the private sector to undertake road maintenance and construction expands, the government road authority should gradually change its role from an executing agency to a contract supervisory agency. This change produces a multitude of problems, primarily caused by the fact that the road authority staff will be accustomed to organising labour groups and not accustomed to managing contracts. They will therefore need to adapt to new procedures and roles, and this implies a further need to restructure the road authority and re-orientate staff at all levels to highlight the new tasks which they will have to undertake when managing road contracts.

Institution building

Specific programmes may be implemented to develop the agency responsible for the management of the road network. These programmes must result in an organisation which is sustainable after the donor funding is withdrawn in order to achieve a long term institutional benefit. This reorientation and retraining programme will include the technical issues of contract administration, but will also require the client's employees to develop a new approach to the work. Contractors who are working on projects with a paid workforce will not tolerate lengthy delays while they are waiting for a response to queries sent to the client organisation.

In the short to medium term it is necessary for the road authority to take on the role of the consulting engineer. It is likely that a partly autonomous group may form within the road authority which undertakes this role of design and contract management. This group may also eventually form the catalyst for the development of a private sector consulting profession. The task of institution building within the road agency is to create an ability and develop that ability to supervise and administer contracts. [1] The development of contract administration capacity will require new staff posts to undertake the different roles. However the downsizing of the direct labour operation is likely to provide sufficient personnel, albeit with retraining, to fill the new roles created by the contract administration section.

It is inadvisable to reduce direct labour capacity too rapidly for two reasons. Firstly, it will take an appreciable time to develop the necessary private contracting capacity. Secondly, there will be some work which is best carried out by direct labour until a full range of period contracts are in operation, such as emergency repairs to roads and structures. There will also be work that is unattractive to private contractors due to the location or the nature of the activity.

Seven stages

The institution building process, although separate from developing the private sector, must be carried out in parallel to the development of the contracting procedures and implementation of private sector-executed construction work. There may be considered to be seven steps in the institution building process. Each of these discrete stages must be undertaken sequentially in order to achieve a successful result, although it is possible for one stage to start before its predecessor has finished.

The 7 Stages of Institution Building

1. Initial contact and research
2. Set up institution building steering committee
3. Initial capacity building
4. Decide on scope of project
5. Establish institutional structures
6. Secondary capacity building
7. Mentorship and monitoring

---

1. Initial contact and research

The first stage is to discuss the issues and receive support from the senior management in the organisation. If the senior management are not supportive or are only likely to pay lip service to the changes, the initiative is likely to be unsuccessful and the project must be reviewed. Research must then be undertaken within the organisation to analyse its level of resources, and to determine its strengths and weaknesses, the extent of the changes required and level of support that will be needed.

2. Institution building steering committee

It is essential that the institutional development process is seen to be 'owned' by the organisation itself and not imposed by an outside body. An institution building steering committee should be set up which will oversee the institutional building and change process. This committee should be accountable to the organisation's director and primarily be made up of members of the organisation. Facilitators involved in the institutional building process may be included in an advisory role, but should not control the committee. The initial task of the committee is to plan the institutional process within the organisation, which will not include the actual development of the contract management programme. The committee should then meet regularly to review the change process, highlight where problems occur and initiate solutions to these problems.

3. Initial capacity building

The first priority in capacity building is to ensure that all the staff of the road authority understand the principles and practice of contract administration and contract procedures. It is not necessary for them at this stage to understand their own roles within the system precisely, but they should have a general understanding of the process. It is likely that this stage could take a considerable time, especially if the road authority is large. However, the steering committee along with facilitators would be able to move on to the fourth institution building stage before this stage had been completed. The most appropriate method for undertaking the initial capacity building at this stage is likely to be a series of seminars aimed at different levels of staff in the road authority.

4. Deciding on project scope

The fourth stage of the institution building process is to reconcile the resource requirements of the contract administration project and future requirements for undertaking contract management with the capacity within the road authority. It is essential at this stage that the contract administration implementation project is fully detailed and designed to ensure that the institutional building process is able to provide the necessary capacity for all aspects of the project. This will be the main period in the institutional building process where close links will have to be made with the implementation project. It essential that this process is carried out thoroughly as the additional requirements highlighted may include the need for additional staff in regional offices to manage contracts or an improved communication procedure between the regional and main offices of the road authority.

5. Establish institutional structures

Following the design of the contract implementation project it will be necessary to define the institutional structures (organisational and tasks) that will be needed to manage and supervise the contracts which will be let by the agency. These are the roles which will be required for the management of the actual contract administration and not the institution building process which is overseen by the separate committee. It will be necessary to define the different roles that will be undertaken by staff members, including job descriptions and tasks involved with each contract administration job. When the institutional structure is being determined, it may be useful to build on the existing skills of different jobs from the existing structure to define new roles that staff can move into. It will also at this stage be necessary to define the terms of reference for the institution building facilitators who will undertake the training of the staff in their new roles.

6. Secondary capacity building

Following the defining of the new roles, it is then necessary to undertake the secondary capacity building by providing the particular skills which are needed by each member of staff. This will include explaining their new role and teaching the skills that they will need to undertake to tasks associated with the role. It is likely that, if there are a large number of staff to be retrained, this stage will have to be repeated a number of times to increase the capacity within the road authority. It may be possible to start undertaking roadworks using contract documentation and the private sector on a limited scale following the training of an initial group of road authority staff, provided some training has been carried out for all the new roles in the road authority. Full implementation of contract documentation will not be possible until all the staff have been trained, which may take many months.

7. Mentorship and monitoring

The initial training that can be provided to each individual staff member will be provided over a short time, usually through a training course. A further period is required for the staff to adapt to their new role and fully understand the tasks which they have to undertake. After the re-training of the initial group of staff, this can best be provided through a mentoring programme where staff who are more experienced in their new roles are able to offer advice and information to staff who have only recently been trained. However, initially it will be necessary for the trainers and facilitators to monitor and support the first batch of staff who are managing contracts, helping them to improve their performance until they are able to work unassisted.

Implementing the process

The institution building process must always be undertaken in parallel with the implementation of the contract administration project. The diagram below shows how the seven steps described in the institution building process fit chronologically with the implementation stages of the contract administration project.

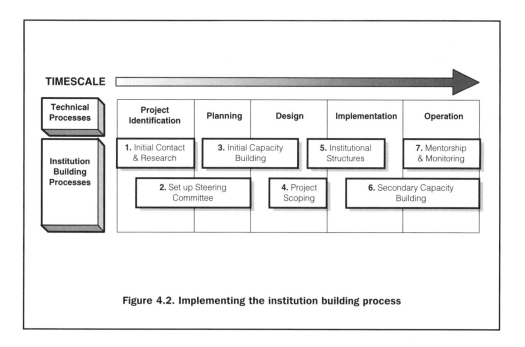

Figure 4.2. Implementing the institution building process

Stages 1 to 6 must be completed before the implementation of work under-taken by the private sector, even on a limited scale. Stage 6 can continue to be undertaken during the management of the first private sector contracts, build-ing the capacity within the road authority to manage the contracting out of all of its road construction and maintenance work. The final stage of the institu-tion building process, mentorship and monitoring, will continue until the road authority and its personnel have settled into their new roles.

Change management

A critical factor in the success in the institution building process will be the willingness of staff to accept the cultural changes required in adapting to their new roles. Previously the staff will have been working in a relatively safe environment with high job security and limited personal accountability. The new contracting approach will require them to take on more responsibility but with reduced job security, as discussed in Chapter 3. The institution building programme should therefore address the issue of change management to assist employees to adapt to the new roles that they will be required to undertake.

Capacity building

The staff of the road agency will have to cope with a range of new and unfamiliar roles. For example, within the contract administration section there will be requirements for:

- contract preparation;
- pre-qualification of contractors;
- tender evaluation;
- project supervision; and
- work monitoring and payment authorisation.

In order to enable the contract administration team to provide the best value for money, they must have a database of the road network that contains information about road condition, level of traffic and existing construction and maintenance undertaken. This will allow the efficient planning and prioritising of maintenance and rehabilitation contracts.

Training and related measures

As discussed in Chapter 3 it is likely that the training, or retraining, of client staff will involve a heavy emphasis on attitudinal change rather than merely developing skills and knowledge. This retraining could be achieved through a mix of coaching, consultancy and participatory workshops. Before the training commences a careful review of the roles for staff in the new road agency should be undertaken. This will highlight the changes that will be needed if the existing staff are to adapt to their new roles, together with measures that will have to be implemented to allow re-deployment of staff who will not have a role in the new organisation. Wherever possible training should be provided for staff to enable them to be re-employed at an equivalent level in the new road authority.

For training to be successful there are five factors which must be addressed:

- training systems should be developed and functional;
- documentation and materials should be available;
- training material should be developed;
- training should be practical and realistic; and
- trainers should be competent.

These are short term measures to enable the client organisation adapt to the new roles that it will need to undertake. It will be necessary to develop longer term training capacity that can provide support to the road authority and industry in general. This need relates not only to technical training but also to construction management which is currently often either poorly taught or completely ignored. An overview of construction management training needs in developing countries is given in Table 4.1 below[2].

Table 4.1. Construction management target groups and their training needs		
Target group	**Typical education background**	**Training needs**
Policy-makers	Various	■ Appreciation of special characteristics of the construction industry ■ Appreciation of industry's development needs

Mangers of construction programmes	Engineers, Architects, Administrators	■ Project management: techniques and general approaches to:
		  – Planning;
		  – Contract preparation and administration; and
		  – Control of construction

Managers of construction operations		
General Managers and owners of construction firms	Professional and/or technical training and/or long practical experience	■ Strategic planning ■ Estimating ■ Tendering ■ Financial management ■ Legal aspects ■ Personnel management ■ Plant operation ■ Purchasing ■ Risk assessment
Site Managers	Technician/technical	■ Site Layout ■ Planning and control of site ■ Documentation and basic costing ■ Plant performance and operation ■ Site safety and welfare
Foremen and Site supervisors	Vocational training plus practical experience	■ Basic site organisation ■ Human relations ■ Site safety
Construction Management trainers	Engineers/Architects with practical experience and aptitude for training	■ Preparation of training packages ■ Course management ■ Training techniques and methodology

Source: Miles 1993.

Education and training institutions
It will probably be necessary to develop the education and training sector to enable it to deliver these improved engineering and new management courses. Possible institutions include:

- universities;
- government agencies;
- training groups within the public sector;
- private training organisations; and
- construction industry development organisations

The types of training that will be offered by these institutions is likely to vary in length from months or years for university courses to one or two day courses provided by private training organisations. This variety of courses should enable the full spectrum of management access to the types of training that they require; short one day courses for busy senior managers, longer courses for middle management and degree or diploma courses for junior posts. It is likely that courses will have to be financed directly by clients who will therefore only undertake courses if they can see direct results. Training establishments must therefore work closely with the industry to prepare relevant courses that provide the skills demanded by the road authority and other client organisations. The evaluation of selected projects in Chapter 6, provides an indication of some alternative approaches and their applicability in specific situations.

Realistic remuneration
A well run road agency, like any other business organisation, requires a highly motivated and well trained workforce. A shortage of suitable staff is often a problem in the efficient running of the road authority. This is exacerbated by the poor salaries and wages that frequently prevail in the public sector, leading to the cynical attitude that 'they pretend to pay us, and we pretend to work'. Road boards must accept that realistic salaries need to be paid to agency staff in order for them to remain motivated in the work they undertake and discourage them from transferring to the private sector.

A reserved market?
In order to assist the development of the labour-based small scale contractors the road agency could reserve particular contracts for this target group. At first sight this option might seem an excellent solution to guaranteeing work and providing the new construction firms with the work security that they require, but it subjects the market to the range of problems with captive contractors

introduced in Chapter 1. Essentially there are three areas of risk associated with this approach:

- ***The need for strict entry criteria for this reserved market:*** which is likely to result in a strong resentment from contractors who do not qualify for the reserved market, especially those who fall just outside the selection criteria.

- ***The system will always be open to abuse and corruption:*** by those who administer the scheme and contractors who 'create' new small scale companies to access the market. The road authority is a public body and is therefore publicly accountable for its actions. Unlike a private client, it should have a demonstrably good reason to reject a lower bid from a contractor who does not qualify for the reserved market in favour of a higher bid from a qualifying contractor.

- ***The need to create independent contractors who are able to compete in an open market:*** if new contractors have access to a reserved market they may find great difficulty in becoming independent and able to compete commercially on the open market.

Preference schemes

A better solution than a reserved market is some form of preference or inducement scheme, which could be gradually phased out for larger contracts or as a small contractor becomes more experienced. These schemes include:

1. Price preferences:

Price preference schemes are easy to administer and allow work to be offered to all contractors, but provides the target group with some competitive advantage by modifying their tender price when comparing it against other contractors. Under a price preference scheme any pre-qualified contractor can submit a bid for work. When the bids are assessed for price, the actual bids received from the target group are reduced by a notional percentage, which may result in a target contractor bid becoming apparently more attractive than a non-target group bid which is actually lower. This percentage could be varied depending on the level of support required and could be reduced for higher priced contracts in order to gradually reduce the support to contractors as they bid for larger contracts. Although the bid price would be reduced for the purpose of comparing tenders, the price paid to the successful contractor would be the original bid price. The table 4.2 offers a possible price preference structure for road maintenance contracts. A price preference system has the

Table 4.2. Possible price preference scheme for road maintenance contractors	
Contract value	Price preference (contractor bid reduced for tender comparison)
Up to $10 000	10%
$10 000 – $25 000	5%
$25 000 – $50 000	3%
$50 000 – $100 000	1.5%
Over $100 000	0%

advantage that it allows all contractors to bid for small contracts and also forces emerging contractors to become more independent as they take on larger contracts. However, the scheme is potentially open to abuse and corruption if contractors who do not fulfil the preference criteria are included in the 'target group'.

2. Vertical contract packaging

Vertical contract packaging was discussed in chapter 3 within the section on procurement policy. Vertical contract packaging enables any general contractor to bid on equal terms for small contracts. The benefit for small contracting businesses is that the volume of these contracts is likely to make them unattractive to larger contractors unless they can obtain two or three consecutive contracts. If vertical contract packaging is adopted larger contractors may be prevented from winning more than one or two contracts to eliminate this possibility.

3. Horizontal contract packaging

Horizontal contract packaging is similar in principle to vertical packaging. However, rather than dividing contracts into 'lengths' the maintenance work is divided into activities. This approach allows larger contractors to bid for more complex work, such as road reconstruction and regravelling, and smaller contractors to bid for the smaller specialist contracts, such as grass cutting, pothole patching and culvert clearing (which may be unattractive to the large contractors). As contractors expand their business they would be able to bid for and undertake more complex work that requires more experience and resources.

A key advantage of the contract packaging approach is a reduced scope for abuse and corruption as all contractors are allowed to bid for any contract. The contract packaging system reduces resentment of the group of 'selected' contractors, and encourages contractors to expand their businesses as the incentive to remain small has been removed. The road authority is able to select the cheapest suitable bid, and does not have to justify selecting a price preference tender.

Simplified tendering procedures

Research has shown that contractors face the greatest number of problems when they are preparing to tender for jobs. The work is often only advertised in the main urban areas, even for rural work, and tender documents have to be collected and returned to the regional urban centre which is often remote from the project site and/or the contractor's base. In some cases anybody who claims to be a contractor is able to obtain the tender documents and submit a bid. While it may be acceptable to only advertise large projects in the main urban areas, there is a need to improve small contractors' access to information about invited tenders for smaller projects. The decentralisation of road authority operations discussed in the previous chapter would facilitate access to information about potential work and tender documents. In the short term tender periods could be extended for small works, enabling rural contractors to obtain information about potential work, collect tender documents, prepare and return their bids to the central office.

In order to reduce the work load and cost of assessing tender submissions, a contractor classification and prequalification system should be introduced to prevent frivolous and unqualified contractors bidding for work. Contractor classification systems were discussed in chapter 3 and prequalification systems are discussed in the section on contractors below.

Bill of quantities

The majority of contracts are offered using an admeasurement contract, more commonly known as a bill of quantities (BoQ). Completing the bill of quantities appears to be where contractors experience the majority of their problems. Within the BoQ the total job is divided into specific work items which means that contractors will have to sort through the BoQ to find all the work items that require a certain material and artisan. For example it will be necessary to determine the number of work items that require laid bricks and then determine a price for this work based on the price of bricks and other materials, transport and labour costs. These costs will then have to be distributed between all the work items that involve brick laying.

Materials and labour list

For small projects a simplification of this system would be to produce a materials and labour list from the BoQ, and allow the contractor simply to insert prices into the list, which would include a portion for overheads and profit, rather than have to distribute the costs into the BoQ. This procedure will involve additional work for the client organisation that may not be justifiable on one-off projects. However, for repetitive projects such as health centres and schools the additional cost for each individual project is unlikely to be very great. There are other methods of pricing construction projects that would simplify the tender process, these are discussed in the section on alternative contract payment systems below.

Tender evaluation

Once the tender documents have been submitted, contractors frequently complain that the tender evaluations take a long time to complete and frequently they never hear the final results or receive any feed back on their submission. The restructuring and institution building within the road agency, suggested above, should ensure that tenders are evaluated more swiftly and that the results are published providing some feedback to the contractors on their level of competitiveness.

Alternative contract payment systems

Some of the problems associated with a BoQ contracting system have been discussed above. There are other contract payment systems that can be used by client organisations to obtain a price for the work that they want undertaken.

1. Schedule of rates

A schedule of rates system is similar to a bill of quantities, in that the contractor will return to the client a list of prices for undertaking different items of work. However, it is simpler than the BoQ system for the contractor, as it is not necessary to divide the cost of carrying out work between the different items in a traditional BoQ. The schedule of rates system may be seen as a step between the materials and labour list and the full BoQ. It may also be possible for the client organisation to publish set rates for items that appear in the schedule in order to provide contractors with a guide when preparing their bid. It is likely that this approach would only be adopted for small projects undertaken by inexperienced contractors. The road authority would have to set realistic rates that would enable the contractor to undertake the work to a suitable standard and obtain a reasonable profit. It may be a better option to provide guide rates which the contractor could either exceed or undercut.

2. Cost plus fixed fee

This contract will pay a contractor the costs that are incurred in undertaking the work plus a fee that the contractor defines in the bid. An alternative is for the contractor to be paid a percentage of the costs incurred instead of the fixed fee. This contracting system is probably the simplest for contractors to bid for and for road authorities to evaluate tenders. However, the road authority will have to mobilise substantial supervisory resources during the execution of the work to verify the work undertaken by the contractor. It is clear that this contracting system is very susceptible to corruption, and may be seen by unscrupulous contractors as 'a license to print money. '

3. Lump sum

A lump sum contract is a fixed price contract where the contractor agrees to undertake the work for a fixed price. The advantage of this contract is that the client is assured of the price to complete the project as many of the risks are transferred to the contractor. Although this type of contract appears highly desirable to the road authority, its use is likely to be disastrous for small contractors who underestimate the risks that they will be required to cover.

Table 4.3 summarises the advantages and disadvantages of the three types of contract discussed above. Although the lump sum and cost plus contracts may appear attractive, their disadvantages outweigh the advantages. The BoQ, schedule of rates or materials and labour list systems are the only realistic methods of contract pricing.

Mobilisation payments

Contractors face a large financial commitment at the beginning of their work, including setting up a work site and camp, procuring initial materials, tools and equipment and paying the first wages for their labourers and staff. This occurs before they receive their first payment. The issue of providing mobilisation payments in a contract should not be overlooked as a long term solution to assisting the cash flow of small contractors and reducing the overall cost of the project.

Contract provisions

Many existing contracts do not state the time period for payment to contractors after they have submitted their invoice. If a time period is stated it may often be ignored. To ensure that there is a mechanism for realistic remuneration for contractors, the contract document should clearly state how the following issues will be dealt with and the client should ensure that the contract documents are followed.

Table 4.3. Alternative pricing methods		
Pricing method	**Advantages**	**Disadvantages**
Schedule of rates/ BoQ	■ Simplified schedule systems are easier for contractors to complete ■ Payments are only made for work done ■ Work can be altered and repriced using the tender rates	■ Effective supervision required by road authority to measure work done
Cost plus fixed fee/percentage	■ Provide flexibility to modify the work to be undertaken - e.g. emergency work ■ Simplified tender procedures for contractor and road authority	■ Very tight supervision required by road authority to confirm cost of work completed ■ System open to abuse and corruption
Lump sum	■ Price is fixed at the start of the project	■ All risks transferred to the contractor ■ Contractors may rush work to increase profits ■ Very tight supervision required to ensure quality control

Adverse physical conditions: The contract should make a provision for additional contract time or payments to be made to contractors if the work is more difficult than that which would be reasonably expected.

Price fluctuations: When a contractor bids for work, it is often difficult to predict changes in the price of materials. The contract should therefore permit the rates for each item of work to be adjusted to take account of changes in the price of materials. A price fluctuation clause will protect the contractor against making a large loss if materials prices increase dramatically, and may yield benefits to the client in the form of lower bid prices. Occasionally materials prices will decrease following a shortage and price fluctuation clauses will allow the client to make a saving on the contract price when this situation occurs.

Late payments: If the contractor is late in finishing the contract liquidated damages will be incurred, a charge deducted from the final payment based on the number of days which the contractor is late completing the work. It is normally related to the prevailing interest rates. There is rarely an equitable agreement in a contract to compensate a contractor if payments from the client are delayed. Late payments should be subject to an additional payment to compensate the contractor.

Simplified contract procedures

Many contractor development programmes have experienced problems with inappropriate contract documents, resulting in some programmes developing their own contract documentation. While these 'tailor-made' documents may work well in the short term, they risk two long term problems. Firstly, the contract may not 'stand up' to legal criticism, if it has not been prepared by specialists and conflicts with other local laws and procedures. This may not emerge in the early stages as neophyte contractors would not wish to 'rock the boat' and endanger their part in the programme, but it could store up problems if a serious dispute ensues. Secondly, the programmes usually aim to develop the contractors in order to allow them to enter the open construction market. The programme could therefore not be considered a success if contractors were not familiar with standard contract documentation and procedures. If this approach is adopted there must be a process for 'weaning contractors off the simplified contract' and enabling them to compete using the national contract documents.

Contractor development programmes have sometimes sought to simplify existing contracts by deferring application of additional parts and risks of the original contract documentation until the emerging contractors become more experienced. The differing levels of contract approach was described as a tiered bidding structure in chapter 3. If this approach is adopted a clearly defined set of criteria must be established to determine when a contractor can or should move on to a higher level. In addition, if contractors are prevented from bidding lower than their 'level', newer less experienced contractors would be protected from the competition of the more experienced contractor for their first few contracts. The final level of this tiered system should be a contractor able to undertake contracts with the country's standard contract documentation. Apart from the contract value increasing through the levels, other items could be altered to assist new contractors, such as:

- level of surety required;
- percentage mobilisation payment granted;
- progressive use of schedule of rates, target rates and Bill of Quantities as contractors grow in experience;

- level of technical assistance available;
- taxes and levies on staff, labourers and equipment;
- access to loans at preferential rates;
- removal of certain risks from the contractor (e. g. unforeseen weather, undetected errors in drawings); and
- reduced penalties for late completion.

Although the list is not exhaustive, the items above should not all be included, the choice being dependant on the type of contract applying in a specific country and a careful assessment of the current constraints affecting local contractors.

Equitable conditions

Within the context of developing small contractors, simplified contracts can often be synonymous with equitable contract documents. The client organisation is ultimately responsible for the contract conditions governing the work that a contractor undertakes. Many different organisations may offer advice on the terms and conditions of the contract to achieve a workable document. The client organisation may be tempted to transfer all the risks to the contractor but will ultimately suffer from contractors who are unable to fully evaluate the risks, ignore some risks when bidding to remain competitive or become bankrupt when they are not fully covered against the risks. An equitable contracting system is required which clearly defines the responsibilities of the client in addition to those of the contractor.

The client organisation should be encouraged to look upon the contractor as a partner in achieving the same ultimate goal. Contracts should clearly define the roles and responsibilities of employers and contractors and methods for dealing with issues that include:

- provision for price fluctuations;
- protection against unforeseen ground conditions and adverse physical conditions;
- compensation for late payments;
- realistic level of performance bonds and retention money;
- on-site supervision; and
- vague or ambiguous specifications.

The box below highlights some relevant lessons learnt from World Bank experience.[3]

Lessons from World Bank experience

1. The cost-plus-fee contract has been used in Brazil for implementing road maintenance by contract. It was thought at the time that contractors would not accept any other type of contract because they were not familiar with the risks involved in road maintenance. As always with cost-plus, there was an incentive to inflate inputs and costs, so as to inflate profit. It took 15 years to shift to contracts based on unit prices, once it was realised that cost plus contracts are not much more efficient than direct labour (force account).

2. Grass cutting can easily be checked when the works are completed, but the contract must specify the number of cuts per year and the width of cut. Otherwise cuts will be rare and narrow if paid by lump sum or frequent and broad if paid per square metre (examples from Malaysia and Nigeria).

3. In Chile there are two unit prices per square metre for patching pot holes, one for pavement repairs and one for base and pavement repairs. The bidding documents contain an estimate of the number of pot holes and the area of affected surface, as a basis for the contractor to calculate average unit prices.

4. In Pakistan, the highway authority has a budget for emergency operations, and every year they select three contractors in each zone for these works. A standard format for the bill of quantities is prepared in advance. When an emergency occurs, quantities of itemised works are estimated by the highway engineer, and bids are sought from three contractors with immediate award and issuance of the order to proceed. Actual quantities are measured for payment.

5. Pot hole patching is contracted out on a lump sum basis in the Seychelles, so as to give contractors an incentive to intervene early and patch emerging holes before they grow bigger.

6. Three year contracts for maintenance attached to contracts for rehabilitation are being prepared in Zaire. Maintenance is easier to contract out once the road has been put back in good shape, and the continuing responsibility for maintenance is a positive incentive to the rehabilitation contractor.

Source: Lantran, 1991.

Quality assurance and accountability

The road authority is publicly accountable and must therefore ensure that the work that it undertakes is in the best interest of the public. It must also be able to demonstrate that it utilises the limited resources available in the most efficient manner. In order to achieve the most efficient use of public funds, it will be necessary for the road authority to maintain a road register as described in the management framework section of the previous chapter. A road register, that covers the whole road network, would contain information about the type and construction of the road, the maintenance previously carried out, level of traffic and any defects or problems. This road register will allow the road authority engineers to prioritise the work to be undertaken and make the most effective use of funds available.

When work is commissioned there must be clear specifications available to the contractor detailing the standard of road reconstruction to be provided. If these specifications are to be useful to both the contractor and supervising engineer they must be:

- *Simple:* the specification must be easy to understand and not open to mis-interpretation by the contractor or supervising engineer;

- *Realistic:* the specification must be achievable and easily replicatable with the equipment and expertise that would be expected from a contractor; and

- *Verifiable:* it should be possible to undertake tests and checks with minimal equipment either on site or at a local laboratory.

Safety

The road authority is responsible for the safety of the public both during construction or maintenance of the road and its subsequent use. While it is possible for supervisors to make recommendations or demands of the contractor to undertake work in a safe manner to prevent injury to the public or the work force, the road authority should also consider the safety aspects of the road in use. Specification and design standards should be prepared bearing in mind the needs of the road users, particularly pedestrians. In particular, the road authority should ensure that roads are built with suitable verges for pedestrians and that road alignments do not themselves contribute to increasing risk.

Quality assurance

The road authority should develop a quality assurance system to ensure that a high standard of service is provided to the public. A quality assurance system is distinct from a quality control system that a contractor may operate. For example, under a quality control system a contractor may be regravelling a section of road using three passes of a roller. After the third pass it may be a contract requirement to carry out a compaction test on a section of road and the work would be rejected if it is not up to standard. A quality control system measures an output and rejects items that do not comply. On the other hand a quality assurance system develops procedures that ensure compliance with the standard or specification in the first instance. The road authority should establish a series of procedures for the activities that it undertakes to ensure that it is able to achieve its required goals or outputs, which should be available to all the personnel in the road department.

International information exchange

During the restructuring and capacity building within the road authority it will be necessary to provide mentorship support to personnel. While this support may be provided through facilitators and from within the road authority, additional longer term support may be achieved through international information exchange. For example the World Road Association (PIARC) operates a technical enquiry service known as the World Interchange Network (WIN), and also promotes sharing of information between road agencies in its member countries.

4.3 Contractors

Promoting Contractors' Associations

Contractors, like farmers, are notorious for complaining about the problems that they have to face. While some of these problems are a 'fact of life' associated with operating any business, others are well justified. The main problem that small contractors face is that individually they have little power to bring about changes to their current situation. Each contractor is a small fish in a very big sea. A Contractors' Association could offer a link between the individual contractors and the road authority and other government departments, providing one voice to the government representing contractors and a dissemination route for information and feedback to contractors from government departments. Figure 4.3 highlights the roles that a contractors' association can play in the development of contractors and the construction industry.

Contractors' Associations are able to provide assistance to their membership by offering information services and training. They can also administer a registration and classification system of the kind proposed in Chapter 3,

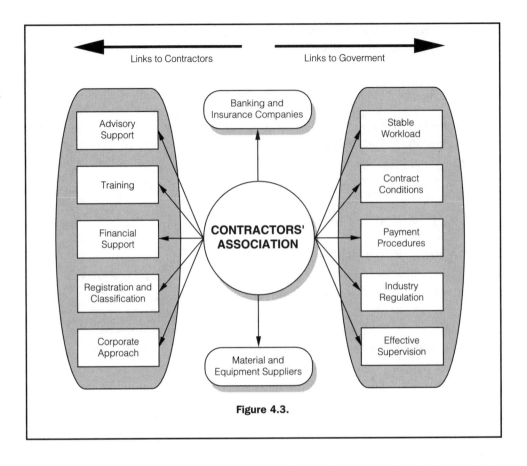

Figure 4.3.

provided that this is acceptable to clients. By acting as one body through the Contractors' Association, it may be possible for contractors to promote a corporate approach to solving the problems faced by the industry. The association can also foster links with banks and suppliers in order to facilitate contractors' access to finance, material and equipment resources. As the contractors' association will be representing the whole of the sector it will 'carry more weight' when negotiating with the government and will therefore be able to offer proposals for improving contract conditions and payment procedures. The association would also be able to offer policy advice, which would hopefully result in a more steady flow of work to its members.

There are four main issues that must be addressed when planning and developing a contractors association:

1. Government acceptance
2. Funding
3. Membership
4. Leadership and management

1. Government acceptance

To be successful the Association must be accepted by the Government as representative of its membership group. However, the government must accept that the organisation is an autonomous body which is set up to represent its membership and therefore may at times criticise government policies and legislation.

2. Funding

The primary problem in setting up a contractors' association is securing enough funding to commence operation and then ensuring a sufficient income to provide a sustainable and realistic level of service to the membership. The are likely to be four main sources of finance:

Membership fees: All members of the association should be required to pay a membership fee. The level of this fee may be fixed or set in bands according to the size of the contracting company, but ultimately would be determined by the perceived level of service provided by the association to its members.

User fees: Certain services, such as information support, could be provided on a fee for use basis, possibly with a certain amount of use included in the membership fee. This charge would not necessarily fully cover the costs for the use of the service, but would assist in meeting the total cost. Charging for these services will also provide members with a perceived value and hence are more likely to be accepted as proper, useful and worthwhile information.

Government support: If the government feels that the contractors' association will assist them in their task of liaison with the industry, then they may initially fund the establishment of the organisation. If the government departments perceive a continued benefit in supporting the association they may provide operational funding, but this may be tied or provided with conditions which will prevent the organisation being autonomous and controlled by its membership.

Donor grants: International donor agencies may provide funding to cover the initial costs of the association in the form of a technical assistance project. This source of funding should only be considered as a short term source and not relied upon for the sustainable operation of the association.

3. Membership

The main issue to be addressed under this heading is should membership be voluntary or compulsory for small contractors? They may be reluctant to participate if they will have to pay a subscription, but it would be difficult to legislate for and police a compulsory membership system (apart from its inevitable unpopularity). The ideal solution would be for membership to be voluntary, but the perceived benefits being great enough for contractors to want to join. This is likely to be achieved if the contractors' association runs an effective contractor classification and registration scheme. If the scheme was supported by the main client organisations (government departments) who would only offer work to registered contractors and membership of the association was necessary to be a registered contractor, the membership issue can be solved (although it could be argued that this is a modified form of compulsion). The registration scheme would assist with the formation of links between the government and the individual contractors and assist with the flow and continuity of work providing that it was effectively managed.

4. Leadership and management

All new organisations require a 'champion' to undertake their initiation and initial operation. This leader will be prepared to work hard to make the association a success and provide it with good foundations for future operation. If there is no obvious potential champion available, it is unlikely that the establishment of a contractors' association will be a success. This champion should be supported by a group of managers who are also committed to the ethos of the organisation. They must be skilled in the tasks that they are required to undertake, in order to ensure that the initial activities of the organisation are a success and to develop a sustainable and effective management structure.

Contractor prequalification

The ideas of contractor classification and prequalification have already been introduced, and they aim to protect the client from inexperienced contractors and reduce the workload of tender evaluation. A prequalification system effectively prevents contractors bidding for work if they do not meet certain criteria, such as the required skills and experience. Additional requirements may relate to other criteria such as level of turnover or the number of trained supervisors. It should be noted that prequalification and classification systems will only be successful if there is a strict set of criteria to determine at what level contractors can prequalify and how they may rise, or be demoted, to a new level. Prequalification or classification criteria are based on monetary bands designated in local currency, and should be adjusted whenever necessary to take account of inflation rates.

The Tanzania Civil Engineering Contractors' Association (TACECA)

TACECA was founded in 1995 to raise the capacity and capability of local contractors. The membership ranges from large companies to small one person enterprises. Each member pays an annual subscription, according to the type of work undertaken, which is the main source of finance for the Association's activities. The main objective of the Association is to protect the interests of, and foster co-operation between, its members enabling an enhanced participation in all construction programmes. It has a five point strategy:

- Increase its membership to 500 by the year 2000
- To press for the formulation of a construction industry policy
- To co-operate with others to establish an industrial development fund
- To address the training needs of local contractors
- To encourage institutional reforms in all sub-sectors of the industry

Two notable activities are:

1. *Encouraging joint ventures:* In these agreements the large contractor bids for contracts, and passes some parts of the work to the small contractors in the joint venture. In some cases the large contractor may also provide construction materials and equipment. The small contractors are paid for the work they have undertaken less a deduction for overheads (approximately 7%) and costs of any materials or equipment provided. The small contractors are effectively provided with the support of a banking system, equipment and material supplies and training.

2. *A proposal to establish a construction industry development fund* (in partnership with the National Construction Council of Tanzania): to provide consultants and contractors with access to funds for working capital and procurement of tools and equipment. It is proposed that the fund will be established by a combination of grants from the government and donor agencies and shares bought by contractors and other stakeholders in the community.

Training and related measures

The construction industry undertakes a large number of activities that requires a broad range of skills, both management and artisan. Table 4.1 above highlights the management skills required by construction managers and supervisors. Many of these skills are learnt on-the-job. This training may be supplemented with more formal in-company training for the staff of large contractors. However, small contractors do not have the resources to provide the additional training courses that they require. Training falls into two categories; formal specific training programmes supported by donor agencies or government departments and informal training provided through ad hoc arrangements made by contractors themselves.

Formal training programmes

Some training programmes are provided at minimal cost to contractors, supported entirely by governments or donor agencies, which makes them very popular and oversubscribed. Whatever selection criteria are chosen, the selection process should be open and transparent. In some training programmes, advertisements in the press and on local radio have been successful in informing potential programme participants. Generally the selection has been undertaken using a questionnaire with a ranking system followed by an interview. Regardless of the financial commitment required from contractors it is important that the course participants are aware of the training that they may expect and the skills that they will have acquired at the end of the course. Contractors may already have ongoing jobs and releasing supervisors to participate in training courses still represents a risk and investment.

As there are many different activities and skills required in the construction industry, the most appropriate training delivery approach is likely to be through a modular programme. Each module could cover a different skill, for example site planning, tender preparation or financial management. There are five stages to preparing a training programme.

1. Survey of contractors training needs
2. Assessment of existing training material
3. Development of modular training material
4. Technical and managerial support to prepare trainers
5. Promotion of training facilities

In the development of the modular training material it will be necessary to review the different training styles that can be adopted (see table 4.4).

Table 4.4. Training styles

Approach	Description
Subject learning	Learning a particular subject through lectures, group work, exercises and discussions, making use of the participants previous knowledge.
Project work	Project work can be used to build on and put into practice skills learnt in subject learning sessions by providing topics to investigate and provide management solutions.
Action learning	This style of learning requires participants to devise and develop a plan or procedure for undertaking a particular activity, such as tender preparation.
Demonstrations	Following the teaching of basic theory and principles it will often be necessary to demonstrate the practices of a topic either on site or classroom demonstrations.
Site practice	This mode of learning allows participants to consolidate their theoretical knowledge by practising the skills that they have previously learnt.

Different topics will require the application of different proportions of each training style to achieve the optimum learning environment. There are a number of other issues that need to be addressed when planning a training programme; including location, training period, choice of individual trainers and provision for follow-up (see table 4.5).

Table 4.5. Issues when planning a training programme

Issue	Considerations
Training location	Will the training be centrally located, requiring contractors to travel to the training centre, or spread out in the regions?
Training periods	How long will each training session last - a number of short sessions or concentrated in one block? The choice is likely to depend on training location and the length of time the participants are able to leave their work place.
Trainers	In addition to being able to prepare and teach training modules, trainers must have a general understanding of the operation of the industry and knowledge of the problems and constraints that contractors work under.
Follow-up	Training course may be relatively short before contractors return or enter the open market. It is likely that they will have 'follow up' questions that need to be dealt with to consolidate their learning. Consideration must therefore be given to short follow up seminars or workshops

Informal training programmes

Much construction industry training is carried out on-the-job, particularly for vocational topics. This is sometimes effective, but it is more likely to be successful when linked to formal apprenticeship schemes and generally-accepted qualifications.

Equipment issues

A MART survey into the use and problems associated with the use of intermediate equipment showed that there is a general lack of awareness by road authorities and contractors regarding the availability, capabilities, flexibility, capital and operating costs, and procurement sources for intermediate equipment. This was exacerbated by the poor awareness among client and contracting organisations of the real cost of owning and operating items of equipment. Many seem unaware that the cost of owning equipment should include loan interest and depreciation in addition to direct running costs (such as operator wages and fuel). The survey also highlighted the lack of proven designs for items of intermediate equipment and inadequate training material in the procurement, management and operation of equipment in general. Contractors were generally unable to hire equipment and experienced great difficulties in obtaining loans to purchase equipment.

The critical issues that need to be addressed in the use of intermediate equipment by contractors are:

- the lack of awareness of the total cost of using different equipment; and
- poor general awareness of the use, ability and arguments for intermediate equipment.

Selection of appropriate tools and equipment

Contractors are always looking to expand their business, but many contractors do not see labour-based contracting as attractive due to the difficulties of managing very large labour forces. Labour-based contractors would frequently prefer to be equipment-based, since they are unaware of the potential risks associated with large equipment holdings. While it is not possible to prevent contractors expanding their firms and dropping the use of labour-based techniques, it would be useful to stress in any development programme the intermediate steps which can be taken between being a labour-based and heavy equipment contractor. It is likely that contractors may opt to concentrate their work in the intermediate level of contracting, due to the financial resources required for heavy equipment contracting and uncertainty regarding the amount of work available.

Tables 4.6, 4.7 and 4.8 highlight the financial resources required by contractors to undertake work at the labour-based, intermediate equipment and heavy plant contracting levels. The implications are that it is highly unlikely that many small contractors will be able to develop into large plant based contractors. However, the level of work required in routine maintenance and gravel road reconstruction in many countries could provide sufficient work for labour- and intermediate equipment-based contractors for many years.

Table 4.6. Routine maintenance labour contractor capital costs		
	Cost (US$ Equivalent)	
	For 100 km of road	For 150 km of road
SECONDHAND PICK-UP Assumed life of 4 years	8,000	8,000
BICYCLES @ 1 per 15 km Each cost US$ 100 Assumed life of 3 years	700	1,000
OFFICE FURNITURE Assumed life of 10 years	450	450
SECOND HAND TYPEWRITER Assumed life of 4 years	200	200
HANDTOOLS STOCK - store & issued (numbers in brackets for 100/150km)		
Hoes (55/80) @US$ 3.5	193	280
Shovels (55/80) @US$ 7.0	385	560
Bush Knives (55/80) @US$ 3.5	193	280
Slashers (55/80) @US$ 3.5	193	280
Rakes (55/80) @US$ 4.0	220	320
Sharpening Files (55/80) @US$ 3.0	165	240
Wheelbarrows (4/6) @US$ 50.0	200	300
Hand Rammers (7/11) @US$ 10	70	110
Culvert Tools (7/11) @US$ 10	70	110
Mason's Hammers (7/11) @US$ 5	35	55
Mattocks (7/11) @US$ 6	42	66
Axes (7/11) @US$ 5	35	55
Crow Bars (2/3) @US$ 6	12	18
Sledge Hammers (2/3) @US$ 10	20	30
Pickaxes (2/3) @US$ 10	20	30
Claw hammers (2/3) @US$ 10	20	30
Tape Measures (7/11) @US$10	70	110
Ditch Templates (7/11) @US$ 10	70	110
Camber Boards (7/11) @US$ 10	70	110
Spirit Levels (7/11) @US$ 10	70	110
Boning Rods (1 set) @US$ 15	15	15
Line & Level (1 set) @US$ 8	8	8
Totals (US$)	11,526	12,877

Source: Intech Associates, 1995 Uganda prices

Table 4.7. Routine maintenance intermediate equipment based contractor capital costs

I - BASIC TRACTOR EQUIPMENT FOR ROUTINE MAINTENANCE

55hp (41kW) 4x2 agricultural tractor	US$22,000
5t fixed body heavy duty trailer	US$6,000
Sub Total	**US$28,000**

Optional 2t towed grader	US$8,000
Optional towed water bowser	US$8,000
Optional pedestrian vibrating roller	US$12,000
TOTAL	**US$56,000**

NOTE: *It is recommended that the optional equipment is hired if possible, particularly where annual utilisation will be low.*

II - BASIC TRACTOR EQUIPMENT FOR EARTH/GRAVEL ROAD RECONSTRUCTION

2 No 100hp(75kW) 4X4 agricultural tractors	US$100,000
2 No 5t heavy towed graders	US$80,000
1 No towed dead-weight roller with transport wheel	US$25,000
1 No towed fuel bowser	US$8,000
1 No towed water bowser	US$8,000
1 No pickup truck	US$20,000
TOTAL	**US$241,000**

NOTE: *Tipper or flat bed trucks can normally be hired for the gravel haulage, with local unskilled labour for quarry development, excavation, loading, (unloading if necessary) and spreading of gravel. This considerably reduces capital investment requirements for the contractor.*

Source: Intech Associates, 1997 Kenya prices

Table 4.8. Routine maintenance and regravelling heavy equipment contractor capital costs

Item	Unit capital cost new# (US$)	No. in fleet	Item capital cost (US$)
Tracked Loading Shovel (Cat 953)#	215,000	1	215,000
Tipper 4x2, 7t ##	55,000	5	275,000
Motorgrader (Cat 140)	250,000	1	250,000
Self propelled Roller	85,000	1	85,000
Fuel Bowser Truck	60,000	1	60,000
Water Bowser Truck	55,000	1	55,000
Service Truck	60,000	1	60,000
Supervision Pick-up	20,000	1	20,000
		Total (US$)	1,020,000

NOTES

#	Compromise, instead of dozer plus wheeled loader
##	10t or 15t tippers would probably be more economic
*	Excludes low loader for plant transportation between sites (typical hire at US$4/km)
**	No allowance for standby items
***	Prices based on typical delivered cost including taxes and duties
****	1997 prices

Source: *Intech Associates, Kenya market prices.*

4.4 Market facilitators

Figure 4.4 illustrates the complexity of the relationships between the stakeholders in the construction industry of client, consultant and contractor, their institutional requirements and the support framework that exists to meet these demands. The lines only indicate relationships which may exist to meet the demands of the construction industry and do not include linkages which may exist for other reasons.

The support framework

Within each sector of the support framework there may be one discrete organisation or number of organisations providing various but similar support to the industry. The organisations within the framework perform their supporting roles in different ways. At one end of the scale some provide a general service to the industry, whilst others are primarily concerned with assisting their own membership.

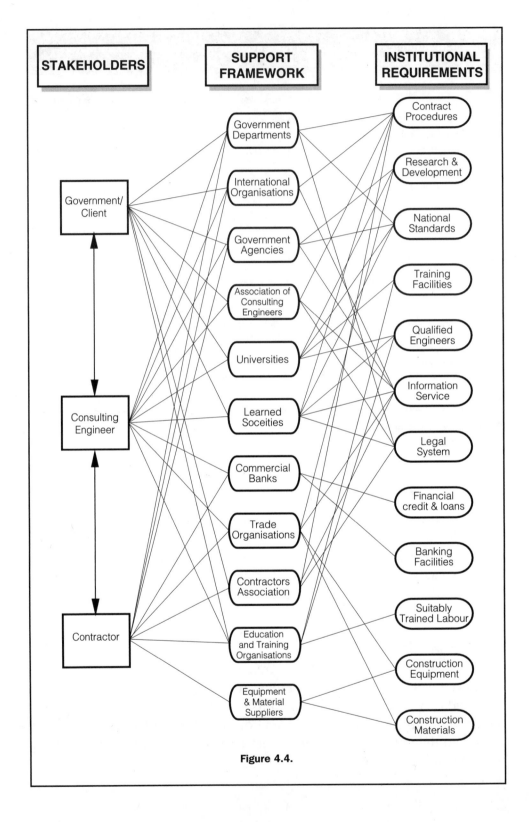

Figure 4.4.

Consultants

In developed countries the majority of contract administration work is undertaken by consulting engineering firms. These firms are able to undertake design work where it is required, prepare contract documents, tender and select contractors and supervise their work and authorise monthly payments. They are in effect acting as agents to the road authority, and undertaking a large part of the work which would have to be carried out by staff directly employed by the road authority. Although the work they are able to undertake will reduce the capacity requirements in the road authority, it does not eliminate the need for an understanding of contract procedures. Unfortunately the consulting engineering profession is very poorly developed in most low income countries. Civil engineering consultants are usually limited to a few small businesses which are run by retired senior officials from the road authority. At present it is easier and quicker to provide the necessary experience and training to enable the road authority to undertake its own contract administration, and developing the engineering consulting profession to undertake contract management will take time.

Autonomous agencies

The World Bank and other influential donors/financing agencies have been attracted to channelling funds through autonomous agencies in recent years. Their primary goal has probably been to improve the prospects of efficient and effective project execution, but the creation of employment and construction industry development have been important associated goals. The model for these agencies is the AGETIP (Agence d'exécution des travaux d'intérêt public contre le sous-emploi), which works largely with and through the private sector. Its tasks include preparation of bidding documents and inspection in the broad sense of "owner's delegate" (in French "mission de maitrise d'ouvrage deleguée").[4] The AGETIP in Senegal arose in 1989 in the wake of concerns related to the social effects of the structural adjustment programme, and it was established as a not-for-profit non-governmental organisation with the following objectives:

- to create employment, particularly in urban areas;

- to provide vocational training, to improve the operational efficiency of the local construction industry and the effectiveness of public institutions;

- to demonstrate the scope for increased application of employment-intensive construction technologies, and

- to execute public works that are worthwhile in both economic and social terms.

AGETIP - Agence d'exécution des travaux d'intérêt public contre le sous-emploi

The association AGETIP, in Senegal, has been given a mission of owner's delegate for a program of small- and medium size labour-based public works and services. AGETIP contracts out all engineer's duties for preparation and supervision to local consultants. Works and services are contracted out to artisans and small- and medium-size contractors. AGETIP carries out the whole management of the project, including inspection tasks. In 1990 AGETIP managed a program of about 130 contracts for works (average value US$ 80,000) and the same number of contracts for consulting services. Standard and computer aided procedures allow AGETIP to pay for works and services within a week.

Source: Lantran, 1991.

The AGETIP approach depends for its success on by-passing cumbersome and bureaucratic government procedures, paying competitive salaries to a comparatively small number of well motivated national staff of high calibre and making extensive use of the private sector. The sustainability of the approach is yet to be proved, and there is a danger that the objectives of the organisation may become excessively diffuse and its independence may be compromised. Nevertheless, it is a positive response to the "feast and famine" nature of demand for construction which is a distinct constraint on poorly capitalised small enterprises, and it can also operate efficient payment procedures so that contractors can plan their work more effectively and gain a reputation for responsible financial management.

The measures for prompt settlement of certified payments due to contractors constitute a major benefit, and help to overcome the widespread demand by contractors for improved access to finance for short term working capital as well as longer term financing of physical assets. The AGETIP system, by paying contractors promptly, avoids the need for direct financing arrangements, a risky business for agencies lacking real banking experience, as careful thought needs to be given to setting appropriate interest rates and collateral requirements.

AGETIP-type agencies have been replicated in a number of countries (notably Benin, Burkina Faso, Mali, Mauritania and Niger), and are likely to continue to demonstrate success while they benefit from a significant flow of external resources. It is, however, difficult for a single organisation to combine development activities with project execution imperatives and conflicting pressures may emerge. Their long term sustainability must also be questioned for two reasons:

1. They are directly financed by external funding agencies are therefore dependant on this funding for their continued operation.

2. Their very nature of bypassing cumbersome and bureaucratic government departments and acting as "owners delegate" managing every aspect of the project will not encourage the necessary changes in public organisations so that they are able to manage infrastructure projects effectively.

Trade organisations

Within the support framework trade organisations are set up to support their membership, particularly material suppliers and manufacturers. Within the UK there are over 180 different trade organisations representing manufacturers of materials including bricks, cement and steel pipes. Each organisation is typically financed by its members and works to promote its own members' interests. This may include promoting the use of the material to contractors and consultants, providing a technical enquiry service and assisting in research and development.

Business associations

Organisations which are comparable to trade organisations are contractor's and consulting engineers' associations. These bodies are also financed by and represent the interests of their members. They can assist in the development of contract procedures and national standards and lobby government on issues which concern their membership. They can also provide advisory services to their members, which can cover technical, financial or legal issues. Both the trade organisations and contracting and consulting associations are able to represent the collective interests of their members and act as a one voice link between the industry and government policy makers. This is of great benefit to their individual members who would have great difficulty individually influencing government policy.

Learned societies

Learned societies, although supported by their membership, take a more global view of the industry with the aim of advancing knowledge and dissemination of information to the whole of the industry. They typically examine and

certify engineers and other technical staff and aim to maintain a high professional competence of their membership. As a focus of engineering knowledge
and experience they can often be called upon to advise the government on
issues relating to the industry.

Education and training institutions

Education and training can be provided by various organisations. At the
highest level universities are able to offer courses for engineers which can
lead to professional accreditation by a learned society. While training organisations and technical colleges offer a range of vocational courses teach the
various construction and business management skills necessary to run a
construction business. Universities are also able to undertake research and
development which can be supported by the industry through sponsorship
from engineering and contracting firms or material manufacturers. Alternatively, grants can also be awarded to universities from government funding
bodies.

Government agencies

Governments can also set up and fund their own agencies to undertake
research and provide information and advice to policy makers. These agencies
typically each work in a specific field of engineering or construction. In the
UK there are a number of these agencies such as the Transport Research
Laboratory (TRL), which provides technical and scientific information to help
formulate and implement government policies and it also carries out research
and related activities in highway engineering and other related topics. As the
funding of these agencies represents a significant investment the government,
the UK has sought to reduce its own financial commitments by privatising
these agencies. The government will still commission specific items of work
from the agencies but it will not finance the whole agency's running costs. It is
therefore necessary for the agencies to obtain additional work from other
sources and build a balanced client portfolio.

A stable framework

The construction industry framework is well established in industrialised
countries. It has evolved over the last two hundred years to meet the needs of
the stakeholders and public demand in general. The tripartite system exists
with each group standing on reasonably equal terms, as contracting and
consulting engineering are well established professions that each have sufficient resources and experience to reason and negotiate with the government.
The risks involved with construction projects are shared equally between the
three groups who each have a similar ability to support them. There are also

sufficient financial resources within the construction sector to fund the support framework even during periods of recession. The challenge for those working in construction industry development programmes is to ensure that their separate outputs contribute to a growing stability in the countries in which they work.

References

[1] Lehobo, A. (1998) Transforming the Labour Construction Unit from an Executing to a Contract Supervisory Agency, in P. Larcher (ed), *Labour-Based Road Construction: A state of the art review*, IT Publications, London.

[2] Miles, D. (1982) *Management Training for the Construction Industry in Developing Countries.* Report to the Tenth Session of the ILO Building, Civil Engineering and Public Works Committee, 1982. The report was subsequently published in modified form as 'Training in Construction Management' in the *NICMAR Journal of Construction Management (Bombay)*, Vol 1, No IV. pp

[3] Lantran, J.M. (1991) *Contracts for road maintenance works agreements for works by direct labour*, World Bank, Washington DC, USA.

[4] Calvo, C.M. (1997) *The Institutional and Financial Framework of Rural Transport Infrastructure*, SSATP Working Paper No. 17, World Bank, Washington DC, USA.

Chapter 5

Designing Change

International interest in the privatisation of public works activities has been stimulated by a growing appreciation that the operation of market forces can generate significant savings through improved operational efficiency. In developing countries, the main difficulty is the lack of a resourceful and experienced private sector which could readily be mobilised to meet new market opportunities. This implies a major change process, in which a local public sector monopoly supplier transforms itself into a commissioning and regulatory authority (or a number of such authorities), and a cohort of local contractors gradually emerges to undertake activities previously executed by direct labour.

5.1 Technical co-operation

The goal of technical co-operation projects within the road sector is to initiate reforms to improve the provision and maintenance of the network. Project inputs and activities may involve both the public and the private sector as well as promoting the interests of various groups of road users. These reforms are intended to help to achieve the super-goal of economic, technical and social development. Pilot projects must demonstrate that the changes which have been initiated are suitable for the country and provide an opportunity to test and refine any problems in the project design. If they prove successful, the refined project design then forms a framework for a broader programme for the whole country to initiate the changes and reforms in all regions. Finally external financial and technical support should be withdrawn without reversing the changes and ensuring sustainability of the new institutional frameworks and systems. It is the final stage, the withdrawal of external support, which is the most critical stage in the process as the assistance must be phased out in a manner which supports sustainability. Executing agencies should always work with all the stakeholders associated with a project to ensure its success.

There are many different options available to donors for the development of technical co-operation projects to promote appropriate forms of road construction and maintenance using the private sector. The choice amongst these options will largely depend on the specific objectives of the project. For example the primary objective of a project may be to improve the road network, and the achievement of an increase of national construction capacity may simply be seen as a useful consequence. Alternatively the main objective of the project may be to increase the country's construction capability, with the possible secondary advantage of improving the road network. These objectives will determine the type of programme design which is undertaken and the issues which should be addressed. Expanding on the example above, if an improvement of the road network is the primary objective then the programme is likely to utilise existing established contractors. However, if the intention is to increase the contracting capacity, then the project should address ways of developing new and expanding existing small scale contractors.

The above description is a deliberate oversimplification of the issues involved in project design, but illustrates the fundamental questions which should be raised at the earliest planning stages of a project so that needs assessment and project preparation are thorough both in terms of overall concepts and practical detail. Contractor development is a relatively new field of endeavour, so there is no tried, tested and definitive project implementation formula that can be applied as a matter of routine. Therefore projects must have flexible criteria and approaches and be able to adjust for differing circumstances.

Five tasks

Five tasks should be addressed when designing a programme, although these tasks can be adapted to highlight secondary factors which emerge during the planning stage:

- Define the new role of the government road department and changes that would be required for it to operate as a contract supervisory agency.

- Assess and plan long-term market prospects (i.e. beyond the life of the original project).

- Determine the size of the project. (i.e. targets in terms of contractors trained and employed or road maintained).

- Define role of secondary institutions such as equipment suppliers, financial organisations and trade organisations.

- Decide on training needs and how they are to be met.

The time frame can vary widely depending on the existing national construction institutional framework. The elements of the institutional framework include not only the existing contracting capacity but also the road agency, and consulting profession in country. The level of changes which may be required by the road agency in order for it to adapt to a contract supervisory agency will have a significant effect on the overall time frame of a project or rate of adaptation to a contract work executing system. Depending on the level of changes required the project may adopt one of two different approaches when dealing with the reorientation of road agency. The first option is to work within the framework of the existing authority, but to undertake reorientation training of the staff to enable them to adapt to their new roles. The alternative is to establish a new contract supervisory agency to oversee and manage construction contracts. The advantage of establishing a separate contract supervisory agency to represent the road authority is the reduced time required establish a working system, especially when there is minimum capacity within the road authority itself. The disadvantage of a separate contract supervisory agency is the possible lack of sustainability due to the continued lack of capacity within the residual road authority.

In many countries a decision is made that the role of contract supervision is to be carried out by consultants rather than full-time staff employed by the authority or agency. This usually yields time and cost savings, but in countries where a domestic engineering consultancy industry is not well-established, it

---

Donor conditionalities

In many cases donor funded projects may have procurement terms which require domestic (donor country) sourcing or supplies, goods or services. When a country, like Tanzania, depends heavily on donor funding for implementing construction projects there may be limited opportunity for local businesses to benefit from the available contracts. This will impact negatively on the development of the local construction industry. A lack of knowledge of the procurement procedures on the side of the local firms may act as a barrier to accessing donor funded projects. This means that donor countries may use indirect means of preventing local firms from winning tenders for donor funded projects.

Tanzanian Civil Engineering Contractors Association (TACECA): Workshop on Local Capacity Building, 1997

will be necessary to introduce a project component to develop their capacity. The dangers of neglecting such a component are illustrated by experience from Tanzania set out in the following box.

Types of assistance

External technical assistance is often necessary in order to cope with the dual change process of introducing labour-based technologies and developing purchaser-provider relationships to enable private sector implementation. International technical assistance can provide:[1]

- *A window on the world:* Access to accumulated international experience through the executing agency responsible for delivering the assistance.

- *Fellowships and study tours:* Provision for direct contact through job exchanges and face to face meeting with individuals to create linkages with established institutions in other countries.

- *Equipment and publications:* Direct assistance in the procurement of materials to obtain items for teaching, research and consultancy which may not be readily available in the country concerned.

- *International expertise:* Provision of short or long term foreign specialists to provide counterpart training and improve the capacity of the institution.

Most projects contain a mix of the four components, and the skill of diagnostic project design lies in determining an appropriate mix of the four inputs to suit the particular needs of the new institution and its staff. Where technical assistance in the form of international expertise is required, it is important to ensure that it is properly defined and that provision is made for a transfer of skills so that the system becomes sustainable without continuing resource inputs. In a sense, international expertise should be seen as a last resort and consideration should always be given to short term intermittent inputs rather than posting long term experts (which tends to be cheaper as well as ensuring that the authority of national staff remains intact).

Contractor development projects

The overall goal of contractor development projects is to improve the private sector capacity for undertaking construction and maintenance projects. Besides encouraging entrepreneurship and more effective work practices, it also creates employment as small contractors are likely to adopt labour based techniques providing a large number of jobs for the unskilled workforce.

Projects can also develop trade skills in the labour force and enhance the business and management skills of supervisors and managers.

The majority of contractor development programmes receive some form of international assistance, generally financial inputs, technical assistance or a combination of both. As this assistance is usually for a limited period, the objective should be to achieve a sustainable level where it will continue without external support. In the case of financial assistance this will usually mean that international funding should be used for the initial capital investment which can then be sustained by the host government, independent company etc. possibly by the use of a revolving fund. Alternatively international funding can meet the difference between the settled running costs of the project and the higher costs during the initial stages of a scheme. These higher costs can be due to lower productivity, higher training costs as a result of 'the learning curve' and higher expatriate salaries. This situation implies that the host government or private companies can invest in the project either from the initiation or as international assistance is phased out. International technical assistance usually takes the form of international specialists who advise on the modalities of the project and assist with the initial training (training of trainers). However with some projects technical assistance can be offered directly to the contractors rather than via training courses generally through on-the-job visits and advice.

Selection criteria

While the main goal of contractor development projects is to increase private sector capacity, there are often subsidiary objectives such as improved maintenance of roads, increased rural employment or wealth distribution. As the programme is usually seen as providing an opportunity to gain competitive advantage, it is likely that many more would-be contractors will apply for selection by the programme than are required to provide a reasonable market. Thus the secondary objectives of the programme will determine selection criteria such as:

- preference to existing contractors;
- preference to new companies;
- gender considerations;
- ethnic origin;
- regional preference; or
- level of existing resources.

Whatever selection criteria are chosen, the selection process should be open and transparent. Press advertisements and local radio can be effective in attracting potential programme participants, perhaps followed up with a public meeting at which the programme can be explained in more detail. At this stage, there should be emphasis on the costs, risks and responsibilities that will be faced by the participants, to weed out those who have the misapprehension that contracting is an easy road to riches. Generally the selection has then been undertaken using a questionnaire with a ranking system followed by an interview.

Project design

When designing a contractor development programme, the primary and secondary objectives must be determined and agreed from the start. This will allow needs assessment and project preparation tasks to be thorough both in terms of overall concepts and practical detail. Despite a range of experimental contractor development projects that have been executed over recent years, there is no definitive project implementation model that can be guaranteed to work in all circumstances. Thus the design of such projects remains more of an art than a science, and projects should have sufficient flexibility to cope with unexpected circumstances as they arise.

Since readers rightly expect some guidance, in the following section we offer a model based on direct experience of one of us in project design and management. This is then supported by a series of briefer case studies, which illustrate some alternative approaches. The chapter concludes with some guidelines on project design presented in tabular form.

5.2 A contractor Development Model: ROMAR, Lesotho

ROMAR (standing for road maintenance and regravelling) was a 3-year project to support the privatisation of operational activities related to the construction and maintenance of gravel roads in Lesotho.[2] Lesotho is included in the group of low-income economies in the World Bank's 1994 World Development Report, with GNP per capita of US$590 in 1992 and a negative annual growth of 0.5 per cent over the period 1980-92. With a dispersed population and serious unemployment, it was one of the first countries in the region to appreciate the scope for employment creation through applying labour-based road construction and maintenance technologies.

A Labour Construction Unit (LCU) had been established within the Ministry of Works in 1977, and during the subsequent 15 years grew to a stage where it was recognised as a full branch of the Ministry with 260 support staff and a manual labour force of about 1,800. Besides new construction, by 1992 the LCU carried out routine and periodic maintenance on about 700km of gravel roads, and anticipated that it would be responsible for upgrading, rehabilitation and maintenance of a network of 2,300 km of low-volume gravel roads by the year 2008. However, the shortage of local engineering, technical and managerial staff during a period of rapid expansion, led to heavy dependence on continuing expatriate technical assistance and the development of an excessively hierarchical structure with execution exclusively through force account (direct labour).[3]

The Government recognised the need for a new emphasis on sustainability by securing the involvement of the private sector, so as to reduce the projected government establishment while substantially increasing private sector employment and achieving overall cost savings and greater operational efficiency and flexibility.[4] This implied that the LCU should gradually transform itself into a client organisation which would be responsible for planning, budgeting, design, contract award and supervision, leaving implementation to the domestic contracting industry. The difficulty was that there were no domestic contractors specialising in roads and civil construction, and domestic building contractors were poorly capitalised and lacked essential managerial and business skills. Thus the establishment of private sector companies capable of carrying out labour-based road construction and maintenance was essential, and the acceleration of this change process would require external support in the form of international technical assistance.

The ROMAR project

The priority was to develop capacity in road maintenance and regravelling (known as ROMAR) rather than road construction and rehabilitation (ROCAR) since:

- the immediate need was to maintain and regravel existing roads:

- ROMAR activities offer a more regular workload and source of income for new firms with limited resources and experience;

- experience learned on a ROMAR technical cooperation project could be used to refine a later ROCAR project and thereby enhance the likelihood of success in this more challenging task.

The first step was a study of the domestic contracting industry, which led to the formulation of a proposal for a 30-month technical cooperation project entitled *Entrepreneurship development for labour-based road maintenance contractors*, with the aim of producing an output of 15 trained ROMAR contractors who would be in a position to compete for LCU routine maintenance and regravelling contracts. Accepting a probable attrition rate of 25 per cent during the development process (a realistic figure for pioneering sectoral small enterprise development), the project team was expected to select and train two batches of ten domestic contractors, and also implement various enabling activities such as the development of appropriate contract documents

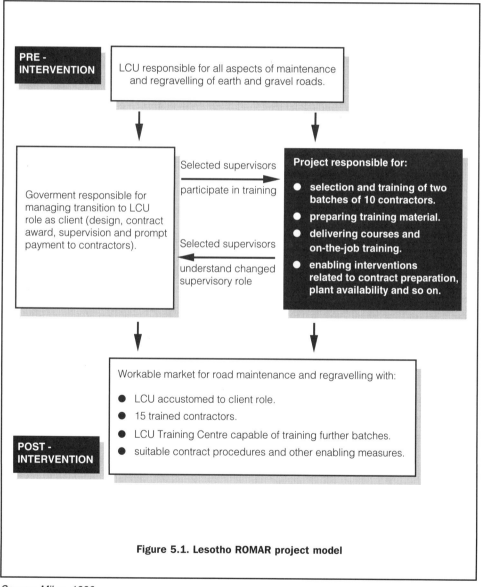

Figure 5.1. Lesotho ROMAR project model

Source: Miles, 1996

and measures to ensure the availability of specialist equipment (notably pedestrian controlled vibrating rollers). The project also developed local training skills so as to permit the training of further batches of prospective contractors as the market expands.

The project model (see Fig. 5.1) demonstrates that the achievement of project objectives depended upon the LCU becoming fully accustomed to a client role, in which it manages the market to which the contractors are offering their services. The project strategy relied upon certain assumptions, notably 'successful integration of force account and contract works in the LCU', but the client felt that it would be able to cope with this organisational change and would not require any significant degree of external support. Thus the only link between the two parts of the process was the provision within the technical cooperation project of six places for LCU supervisors on training courses to be run for the prospective contractors, in the hope that they would gain a better understanding of their changing role.

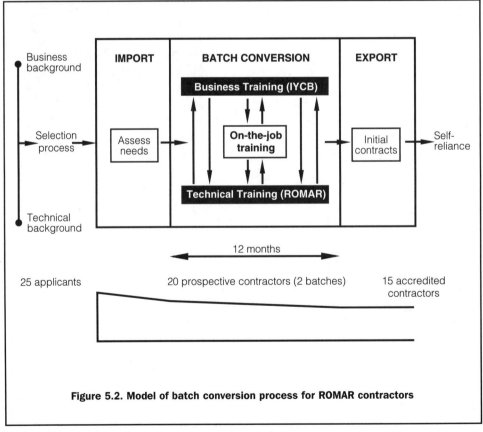

Figure 5.2. Model of batch conversion process for ROMAR contractors

Source: Miles, 1996

The change process

The range of project activities is illustrated in Fig. 5.2. Like all change processes, it consists of three successive stages: *Import* when a group of individuals (in this case a 'batch' of contractors) is selected for the change process, *Conversion* when appropriate attitudes, skills and knowledge are developed, and *Export* when the group is able to cope with their new situation.[5]

Import: The task of selection was particularly difficult in view of the dearth of domestic contractors with public works experience. It was envisaged that applicants would possess either a compatible business background (usually as building contractors) or a technical background with labour-based road building experience, probably as an LCU Senior Technical Officer or Technical Officer. In fact applicants with a commercial background were usually better able to cope with the transition, although the needs assessment (and subsequent project experience) demonstrated that even this group were in considerable need of business training.

Batch conversion: To supplement on-the-job training, the project included classroom training of two kinds:

- Business training, using a standard package of three Handbooks and three Workbooks entitled *Improve Your Construction Business (IYCB)*[6] that had been prepared specifically to meet the needs of owners and managers of small construction enterprises in developing countries; and

- Technical training, using a *Routine Maintenance and Regravelling (ROMAR)* package developed during the implementation of the project.

Export: Following participation in supervised pilot projects and the award of initial contracts, the contractors were expected to be self-reliant given the necessary market conditions (adequate flow of work, payment in accordance with contract provisions and so on).

The accreditation ladder

The nature of the batch conversion process evolved over the course of the project, and was designed to enable participant contractors to climb a ladder leading to full accreditation as ROMAR contractors. Before they could prove themselves on test contracts, they underwent a training process which alternated classroom and practical training so that they could apply and consolidate their skills. Their progress was evaluated at the end of the training sessions, before they were permitted to undertake test contracts. The successive stages were as follows:[7]

- Classroom technical training (three weeks, based on ROMAR).
- Practical field training (four weeks).
- Classroom business training (three weeks, based on IYCB).
- Supervised practical training in contract execution (six weeks).
- Classroom training in understanding contract documents, bidding procedures (one week).

Candidates who successfully completed this stage were permitted to bid for:
- Routine maintenance test contracts (12 weeks).

Candidates who successfully completed this stage were permitted to bid for:
- Regravelling test contracts (12 weeks).

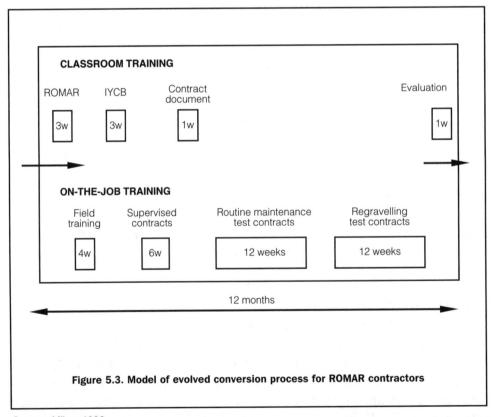

Figure 5.3. Model of evolved conversion process for ROMAR contractors

Source: Miles, 1996

Step-by-step

Following a final evaluation session, successful candidates were recognised as accredited contractors for labour-based road maintenance works. Although there has been some discussion of alternative structures, including an integration of technical and business training, there is considerable evidence that a step-by-step approach to training is usually more successful in skill development among owners and managers of small businesses who do not have recent experience of training in an academic environment.[8] Accordingly the model shown in Fig.5.3 appears well suited for training contractors with relatively little relevant business experience. This procedure might be simplified and shortened where the need for skills development is less substantial.

Output achieved

Overall, the project was successful in delivering trained contractors. The enabling activities were implemented as planned, while the project team responded to demand by accepting a larger number of prospective entrepreneurs for training (see Table 1). As a result the promised project output of 15 trained routine maintenance contractors was significantly exceeded, and the number of accredited regravelling contractors is in line with the LCU's requirements.

Need for institutional change

However, there are indications that the limited attention devoted to changing attitudes and procedures in the client organisation may inhibit the proper functioning of the new market for ROMAR activities. The trained contractors remain commercially fragile and poorly capitalised, and need a reasonably predictable flow of work and regular payments in order to establish themselves as self-sufficient construction enterprises. The project deliberately

Table 5.1. Output of accredited ROMAR contractors				
	Batch1	*Batch 2*	*Batch 3*	*Actual target*
Selected for training	15	12	27	20
Accredited for routine maintenance	12	10	22	15
Accredited for regravelling	8	7	15	15

Source: ROCAR Strategy Note. International Labour Office, Geneva, 1994 (Unpublished)

refrained from offering direct financial support to participant firms or substantial equipment loans, on the grounds that this would be seen as 'easy money' and could deflect the contractors from the priority of becoming more cost conscious and learning to operate within tight internal financial disciplines. The project model shown in table 5.4 below was not available at the time that the project was designed, which may have been a factor in the relative neglect of the need to support the institutional change process within the LCU, and a key lesson of this project experience is the dual nature of the change process and the consequent need to prepare the client organisation as well as the nascent contractors for their respective and complementary roles in the emerging market.

A dual change process

Changing attitudes is more difficult and time consuming than imparting skills and knowledge, but it is essential if changes are to outlast the project inputs. Thus an assessment of the ROMAR project offered the "tentative hypothesis" that, although it has achieved its objectives with regard to enterprise development, the need for the LCU to adapt its operations and procedures in a fundamental way called for more co-ordinated intervention than was foreseen or provided for.[9] This hypothesis led to a proposed strategy for a follow-up ROCAR project (to promote road construction and rehabilitation contractors) which recognises that the dual nature of the conversion process, and the requirement that *both* sub-processes must be successful if the project as a whole is to succeed (see Fig. 5.4).

5.3 Other approaches to contractor development

The Lesotho case study provides a framework for the planning, growth and evaluation of a contractor development programme. Four of the following five short national case studies (covering Ghana, South Africa (2), and Uganda) arose from the MART research programme[10], and a case for Tanzania has also been included[11]). They review other approaches to the design and implementation of contractor development projects with a variety of specific objectives, taking account of the following factors which affect the design of these programmes:

- the role and nature of international assistance;
- training programme;
- equipment provision;
- contract and payment procedures; and
- contractor selection procedures.

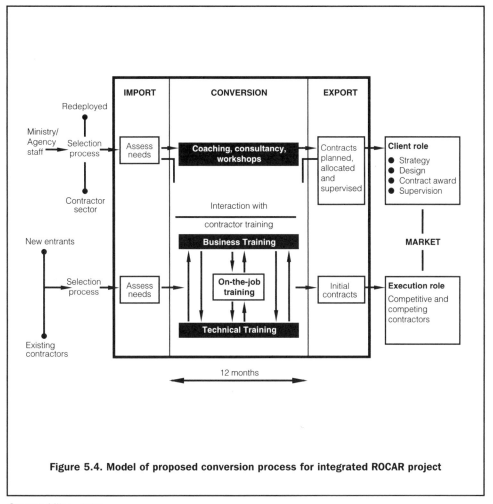

Figure 5.4. Model of proposed conversion process for integrated ROCAR project

Source: Miles, 1996

Ghana: Department of Feeder Roads

The Ghana Department of Feeder Roads has been pursuing a long term Labour-based programme, which is summarised in the following box.

This can be seen as an example of an attempt to create a group of *captive* contractors, who will only be expected to work for that particular agency and who will have a narrow technical specialisation (see Chapter 1). As noted there, the risk is that the new firms will be very vulnerable during periods when the agency is unable to keep up a flow of funded work and are consequently quite likely to fail, with the result that the investment in contrac-

Ghana: Department of Feeder Roads

The project commenced in 1986 and 93 contractors have been trained to work on labour based road rehabilitation and maintenance contracts. The objectives are to:

- improve rural accessibility;
- increase contracting capacity; and
- create rural employment.

The work has mainly been funded by UNDP and the World Bank, and the ILO has been heavily involved as executing agency. The technical achievements have been impressive, and have resulted in the rehabilitation of 1400 km of rural roads at a cost of $14 million. The programme is promoted to contractors by a newspaper advertisement campaign and selection is based on education, previous experience and locality of business. There are three stages to the training process, which addresses the needs of both contractors and DFR staff:

Stage 1.   20 weeks of classroom and fieldwork training
Stage 2.   4 months trial contract of 5 km carried out under supervision
Stage 3.   4 year development with on-site training undertaking a 20 km contract p.a.

Following their period of initial training, the contractors are provided with a set of equipment, consisting of 2 tractors, 4 trailers, 1 water bowser and 2 pedestrian controlled vibrating rollers. The equipment set costs of the order of $150 000, and is financed through a bank loan which is repayable over the following four year development period.[3]

This loan repayment represents a significant element of the contractors' overhead as the bank interest rate in Ghana is about 35%. In order to ensure that contractors are able to repay their loans the Department guarantees contracts will be awarded for the first 4 years after training. Each contract lasts approximately one year and has a value of $240 000. The project attempted to operate these contracts under a competitive tendering system, however, the formation of cartels forced the Department to adopt a schedule of rates for the initial 4 year period. Following the repayment of the equipment loan contractors competed for work through competitive tendering in an open market.

Source: Larcher, 1998

tor any training will be wasted. Fortunately this does not appear to have occurred on great scale in Ghana, but it is a factor to bear in mind before proposing such an approach. The problem is made worse where the agency provides financial guarantees to support equipment acquisition (as in Ghana), since it will be understood that the contractor will only be able to repay outstanding loans if the work continues to flow.

South Africa: Soweto and Winterveld

Following the abolition of apartheid in April 1993 the South African government introduced the Reconstruction and Development Programme which aimed to maximise job creation. Small scale contractors existed in South Africa before, but had usually undertaken labour-only work as subcontractors to larger firms. A variety of projects aimed to develop the employment and business prospects of emerging contractors, including the Soweto Contractor Development Programme and the Winterveld Presidential Project (see following boxes). The objectives of both projects were employment creation, transfer of marketable skills whilst also improving the infrastructure in the area. A common feature of these two programmes was a tiered tendering

---

South Africa: Soweto

The Soweto programme adopted three different approaches for improving the skills of small contractors, ex-supervisors and labourers:

Development Team: The contractor is assigned construction managers, engineers and materials managers who assist with administration of the contract, technical training and the engagement of specialist subcontractors.

Managing Contractor: A large contractor administers the contract while training and supplying materials to a labour-only subcontractor.

Mentorship: This approach is use for more experienced contractors, who employ consultants (mentors) to assist with tender preparation and business management.

Source: Larcher 1998.

South Africa: Winterveld

The Winterveld Presidential Project adopted a more formal approach to the training of contractors, carried out in two phases. The first phase was project specific, enabling contractors to submit realistic bids for the Winterveld contracts. The second phase, which utilises the ILO IYCB training material, was designed to provide the participating contractors with the skills that would be needed to compete in the open market. The Winterveld contract structure is shown below.

Level	Assessment of skills and experience	Maximum contract value in rand ($ 1.00 = R 3.65 – 1994 values)	Performance guarantees
A	Some ability to organise. Limited artisan skill.	Cost of labour component, including contractor's mark-up and profit, to a maximum of R 10,000.	Not required
B	Established artisan. Civil engineering ganger, charge hand, gang boss.	Cost of labour component, including contractor's mark-up and profit, to a maximum of R 40,000.	Not required
C	Advanced gang or trade managerial ability.	Total contract price, to a maximum of R 250,000.	Not required
D	Advanced general management ability. Commercial experience.	Total contract price, to a maximum of R 850,000.	5 per cent of contract price.
E	Advanced construction management ability. Marketing skills. Credibility with financial institutions.	Total contract price, to a maximum of R 2,500,000.	10 per cent of contract price

Source: Larcher 1998.

structure, which was intended to prevent contractors bidding for contracts outside their assessed level of capability and prevent more experienced contractors from dominating the small contract market. Contractors progress to a higher level as they gain experience until they reach the final level, which is that of a national experienced contractor.

Tanzania: labour based road contractor training project

This project commenced in 1992 with two objectives:

- establishing a labour based contracting capacity within two regions of the country
- increasing the capacity of the National Construction Council to take over the operation of the training programme.

Tanzania: labour based road contractor training project

The project is part of an Integrated Road Project financed by the UNDP, IDA, USAID and the Government of Tanzania. In the first three years of the project 12 contractors were trained in each of the two regions in 2 batches of 6 contractors each. All of the contractors selected for training under the project were registered building contractors who each have an annual turnover of $60 000 and employ approximately 70 workers.

Three supervisors from each firm receive 6 weeks classroom teaching followed by 14 weeks fieldwork training. The contractors then undertake 6 month trial contracts to maintain a 5 km section of road. During this period the directors of the contracting firms undertake a course in contract management which aims to improve their business skills.

All the contract work is undertaken with hired equipment which is available on the open market. In order to ensure that the contractor is able to procure the necessary equipment to carry out the work, they receive a mobilisation payment equal to 30% of the contract sum. While 15% goes directly into the contractor's bank account the other 15% is paid directly to a plant hire company as an advance against the plant hire costs.

The project has used a number of different contracts initially utilising FIDIC (3rd Edition) with adaptions for labour based roadworks. However, a new contract has been proposed for use which is based on *The administration of labour -intensive works done by contract* (Garnier and Van Imschoot)

Source: Osei-Bonsu 1995

Uganda: labour based contracting programme

Since 1986 the government of Uganda has rehabilitated 50% of the main roads and 28% of feeder roads at a cost of $300M. With a view to maintaining this investment and developing a local construction industry they looked to small private contractors to carry out routine maintenance of the network.[1] The Labour Based Contracting Programme was started in 1993 and is financed solely by the government who invests $3.2M per annum.

Uganda: labour based contracting programme

The programme is open to anybody with District Engineers selecting suitable candidates under a range of criteria which includes; experience in roadworks, tools and personnel available and a reference from the local 'council' chairman. Contracts are let on a yearly basis with a fixed schedule of rates determined annually. Individual contractors (lengthmen) were awarded 2 km long contracts while small contractors were awarded 10 km long contracts.

The supervision of the contracts was undertaken by the District Engineer and his supervisors. To allow supervisors to adapt to their new roles practice orientated training was provided in contract supervision and quality control. Engineers received training in contract approval and maintenance management, while contractors received on-the-job training in routine maintenance activities and site planning.

Contracts specifically design for the programme implied the use of labour based techniques. The programme initially commenced equipping each contractor with a set of handtools, but was later modified to only providing expensive items such as wheelbarrows. Although the programme is fairly modest when compared to other programmes it has achieved a relatively high output with 8800 km of road maintained in 1996. As the funding for this project has come solely from government funds, payments to contractors are made from decentralised accounts in order to ensure that finance is available each month to pay contractors.

Source: Larcher 1998

5.4 Lessons for future projects

Although the case studies discussed above have widely different approaches the overriding factor which is the backbone for the success of the programme has been the commitment of the government. This commitment and support has had to be available at all levels from central government and the ministry responsible for roads down to the local and regional offices.

Training programme

In some projects technical advice has been offered directly to the contractors rather than via training courses. This has resulted in some cases with the international expert managing the contract or materials arms of the business rather than advising members of the local staff. In these situations contractors have flourished but have then faltered when the support has ended. While there is a role for technical assistance directly to contractors it should be in the form of an advisory role rather than direct input. While each case study has had some form of training element, it should be noted that, while many problems can be attributed to a lack of knowledge of either the contractor and/or the government department, not all problems can be solved by a training programme.

When comparisons of each programme are made it is clear that there is a need for training of government officials as well as contractors. Government officials with experience of monitoring projects which have been carried out by force account have little experience of contract administration procedures and maintenance planning utilising the private sector. Many government officials are also entrenched in the cumbersome force account system. These bureaucratic procedures are not acceptable to small contractors with tight budgets and low working capital. An attitudinal change is also often required in government departments in order for the schemes to operate efficiently and allow contractors to be paid within a reasonable time frame.

When addressing the training needs of the contractor there are various questions which need to be answered, including:

- Who in the contractor's organisation should be trained?
- Should the training be on-the-job or classroom based?
- When should the training be provided?
- What training should be provided i.e. technical / managerial?

- Who should provide the training?
- What structure should the training take?

Variety is inevitable

Inevitably different programmes have addressed the questions in different ways. There are good reasons for this. It is clear that different members of staff in a contractor's organisation require different skills. For example supervisors do not need to be able to prepare contract documents but the ability to plan daily work schedules is essential. Who should receive training would also depend on the size of the contracting organisation. In this book we assume that a small contracting company typically employs about four or five supervisors who report directly to the director. In addition as the director maintains a personal daily supervisory role over each project, it would be appropriate for the manager to undertake the same training as his supervisors. It is unlikely that programmes would be able to undertake training of all a contractor's supervisory staff so the director could be trained in day to day management and technical skills along with (say) three supervisors. The manager should be trained to pass on supervisory skills to existing and future supervisory staff.

Integrating classroom and practical training

Project experience has shown that practical hands-on training in the form of trial contracts has been useful in developing a contractor's skills, however, an element of classroom based training is also essential to analyse and understand the problems encountered on site. Contractors have to interact with other businesses such as banks and insurance companies during the execution of their contracts which often leads to frustrations due to a poor understanding of the constraints that these organisations operate under. Classroom sessions would give the opportunity for workshops/seminars between contractors and representative from these organisations in order for both parties to understand the difficulties experienced by the other. The use of classroom based training sessions implies that training must be carried out on a batch process with a group of contractors who ideally have similar experience in the construction sector.

Figure 5.5 below shows a suggested outline training programme modified from the training programme adopted in Lesotho. A time frame has been deliberately omitted as this would depend upon the number of contractors in the programme and their previous experience.

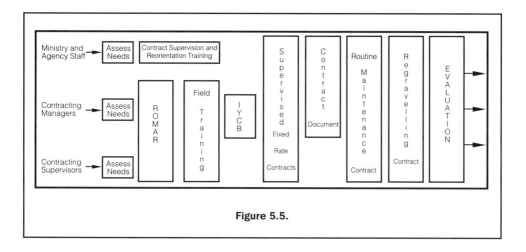

Figure 5.5.

Equipment provision

The solutions to the question of how contractors should be equipped are as numerous as there are projects. Two major factors in the decision of how a contractor should be equipped are:

- The existing level of equipment held by contractors entering the scheme
- The availability on the open market for the purchase or hiring of construction equipment

If a contractor development programme is promoting labour based techniques then it clearly does not need to investigate the issues associated with large civil engineering plant. Existing projects have either given contractors a small quantity of tools and equipment or required them to hire equipment from independent suppliers. The approach adopted has usually depended on the existence of plant hire companies in the country.

Where contractors are provided with equipment it is essential that this is paid for by the contractor. This can usually be achieved during the practical stages on training where the contractor is undertaking maintenance work. The requirement of ownership of tools and equipment results in the contractor taking greater care of his assets and encourages regular maintenance. It also encourages the contractor to maximise the productivity of equipment in order to increase the return on the investment. If the contractor is to be provided with equipment it is essential that it is carefully chosen to ensure reliability and that it is suitable for the work and working procedures that the contractor intends to execute. Rather than forcing contractors to accept a fixed package of tools and equipment the possibility of allowing them to select from a list up to a maximum total value should be considered.

It is likely that the option of the development programme arranging to equip contractors would only be considered where an open free market plant hire sector did not exist in country. This is due to the additional problems for the project associated with administration of equipment loans and financial payments from the contractor. It should also be pointed out that the dangers of setting up a plant hire company within the project where no other companies exist in country are twofold. Firstly the hire company will have a monopoly and hence a strong control over the contractors. Secondly the company invariably ends up being owned or run by the government organisation who also issues the contracts thus increasing the hold over the contractors. In countries where plant hire businesses exist contractors should hire the equipment that they require. The one drawback of this option is that the contractor, who does not own the equipment, is not encouraged to maintain or control its use.

Contracts and payments

The majority of programmes have experienced problems with contract documents, resulting in some programmes developing their own contract documentation. While this approach may work well in the short term, there are two drawbacks. Firstly the contract in some cases would not 'stand up' to legal criticism. This would not pose a problem as the contractors would not wish to 'rock the boat' while they were part of the programme. However if the contract was used for work outside the programme there would be potential for legal cases. Secondly, the programmes were attempting to develop the contractors in order to allow them to enter the open construction market. The programme could therefore not be considered a success if contractors were not familiar with standard contract documentation and procedures.

Other programmes sought to simplify existing contracts including more parts of the original contract documentation as the contractor became more experienced. The differing levels of contract approach adopted by the South African Winterveld Project offered a progressively more testing contract. The principles of more testing levels of contract were outlined in chapters 3 and 4, which indicated that other contract conditions apart from the contract value may be altered for more testing levels of contract. The main contract conditions that may be relaxed for smaller contracts highlighted in chapter 4 were:

- Level of surety required
- % Mobilisation payment granted
- Lowest contract level uses a schedule of rates and lower levels use target rates before a Bill of Quantities is required for higher contract levels

The choice of relaxation of contract conditions will be dependant on the type of contract(s) that a specific country uses. In addition to providing different levels of contract to ensure that there are a realistic number of contracts at each level it may be necessary to slice and package large work into smaller contracts (see chapter 3).

During the preliminary stages of a development programme the issue of prompt payments to contractors must be addressed. For an equipment-based contractor there are two methods available for mitigating the problem of late payment; payments to suppliers can be delayed or part of the working capital can be used to pay urgent bills. However labour-based small contractors do not have these options, since they lack working capital to make payments before they are paid and the majority of their outgoings are in the form of labour wages which must be paid promptly.

Contractor selection

New small road contractors generally appear to come from three main sources:

- Supervisors from larger companies setting up their own business

- Employees from government road departments disillusioned with their current situation

- Small building contractors looking to diversify into new markets.

This range of potential contractors would allow different candidates to be selected to fulfil different roles depending on their past experience, abilities and available resources (see table 5.2 below). In this example those in the first two groups would normally enter as category E contractors in view of their lack of direct entrepreneurial experience. Candidates from the third group would be eligible for entry into group D where their skill and experience would be suitable for culvert construction and repairs to road structures.

This scheme could be linked to the tiered structure discussed above, where contractors with a small amount of experience would not enter the programme at the lowest level and undertake all the training courses, but enter at their specific experience level.

Table 5.2. Private sector maintenance options		
Category	Type of contractor	Work undertaken
E	Labour only contractor	Labour based maintenance work only ■ Vegetation control on verges ■ Culvert clearing ■ Drainage clearing and maintenance
D	Small Scale contractor	Routine Maintenance work ■ Grading ■ Culvert construction and rehabilitation ■ Road Furniture repair ■ Minor repairs to road structures
C	Gravelling contractor	Periodic Maintenance work including regravelling ■ All work listed in category D ■ Regravelling work
B	Large contractor	Major work including construction ■ All work listed in category C ■ New road construction including small structures
A	Large contractor	Major work including construction ■ All work listed in category B ■ Large road schemes including large structures

Source: Intech Associates, 1992, Development of the Private sector: Road maintenance W. Uganda, MoWTC Uganda (unpublished)

Designing a contractor development project

There is no definitive answer or model to the design of a contractor development programme as the situation in each country or region is different. This book aims to highlight and discuss the different issues which should be addressed when utilising the private sector to undertake road construction and maintenance in general. Table 5.3 below provides a summary of the issues that specifically need to be addressed when preparing a contractor development programme.

The table emphasises the importance of the four preparatory tasks which must be undertaken before the final stage of designing the training component. In our experience, most contractor training project failures can be traced to unjustified assumptions and a lack of sensitive organisational analysis and discussion during the period allocated for the preparatory tasks.

Table 5.3. Guidelines for contractor development project design

Task	Key factors
Review current status of client organisation and evaluate assistance that might be required to manage new client role satisfactorily.	■ Determine the likely staff establishment of the agency in its new role, the conversion process that will be needed if the existing staff are to adapt, and prepare a redeployment plan for those individuals for whom a role will not exist in the new organisation. ■ Whenever possible, attempt to secure re-employment at the appropriate level in the new contracting enterprises. ■ Training should emphasise attitudinal change, including a mix of coaching, consultancy and participative workshops.
Determine likely medium/ long term market for contractors.	■ Review agency's plans and budgets. ■ Assess the likelihood of adequate funding being made available to permit these plans to be realised. ■ The number of contractors to be trained will depend upon an assessment of the likely market. Do not aim to develop more contractors than are required to bring about a competitive market, since the result would be 'occasional contractors' without a serious commitment to the industry.
Decide on number of contractors to be promoted, and define training and other needs.	■ Base on market forecast, coupled with a review of resources and experience of existing contractors and their training and related needs. ■ Make realistic allowance for attrition during training process (which will be higher in countries where there is no strong entrepreneurial tradition). ■ Ensure initial selection process is sufficiently testing to screen out those who are clearly unsuitable, and subject classroom and on-the-job training to continuous assessment to expose weaknesses at the earliest possible stage. ■ Remember that the training should help the participants to develop complementary markets so that they can remain competitive and cope with fluctuations in demand from their major client.
Review complementary enabling interventions that may be necessary.	■ Assess current contractual procedures, which may not be suitable or equable for small labour-based contracts. ■ Determine whether existing sources of business finance are sufficient to cover the small contractors' investment and working capital needs. ■ Determine requirements for basic tools and equipment, and whether specialised equipment (e.g. pedestrian-controlled vibrating rollers) can be bought or hired by prospective contractors.
Design training component	■ Review training strategy, including links between business and technical training, the need for step-by-step training and the mix and scheduling of classroom and field training. ■ Review training material to be used in the project, including suitability of published material together with supporting local material either developed or to be developed.

References

[1] Miles, D. and Neale, R. (1991) *Building for tomorrow: international experience in construction industry development*, ILO, Geneva.

[2] Miles, D.W.J. (1996) *Promoting small contractors in Lesotho: Privatisation in Practice. Proceedings of the Institution of Civil Engineers*, 114, Aug., pp 124-129.

[3] Hakengaard A. et al. (1992) 'Report of the Joint Evaluation of the Labour Construction Unit, Ministry of Works, Kingdom of Lesotho'. Swedish International Development Authority, (unpublished).

[4] Project LES/92/02/IDA (1992) 'Description of Services'. Labour Construction Unit, Maseru, Lesotho, (unpublished).

[5] Miller, E.J. and Rice, A.K. (1967) *Systems of organization.* Tavistock Publications, London, pp 33-34.

[6] Andersson, C.A. et al. (1994/1995) *Improve Your Construction Business series: 1. Pricing and bidding, 2. Site management, 3. Business management.* International Labour Office, Geneva.

[7] Engdahl, A. (1994) 'Report on Training Assessment Mission'. (LES/92/02/IDA) ILO/ASIST, Nairobi, (unpublished).

[8] Hernes, T. (1988) *Training Contractors for results: A guide for trainers and training managers.* International Labour Office, Geneva, p 45)

[9] Miles, D. (1995) 'Entrepreneurship Development for Labour-based Road Maintenance Contractors (LES/92/02/IDA)', ROCAR Strategy Note. International Labour Office, Geneva, (unpublished).

[10] Larcher, P. (ed) (1998) *Labour based roadworks: A State of the Art review*, IT Publications, London.

[11] Osei-Bonsu (1995), 'Labour based road rehabilitation and maintenance in Tanzania', ILO, Geneva, (unpublished).

Chapter 6

Implementing Change

6.1 Supportive policies

Effective policies are required to address four issues:

- Appropriate construction technologies.
- A suitable regulatory framework.
- Construction management training needs.
- Constraints on intermediate contractors.

The common thread that runs through this book is the multi-faceted nature of the constraints affecting the intermediate construction sector (small contractors using technologies appropriate for local conditions and resource availability) in low- and middle-income countries. These constraints require definition in order to develop systems and procedures to improve its performance and enable it to reach a sustainable level.

The earlier sections on the concept of appropriate technology and the workings of the construction market in developing countries form an essential foundation for the definition of ways of delivering support and assistance. One key conclusion is that there is a fundamental difference between the operational environment that pertains in developing and industrialised countries, and the corollary is that the institutional framework that governs the industry also needs modification in order to permit an equable and effective role for the target group of intermediate-level construction enterprises.

The first step is to review the special operational environment that pertains in most developing countries. This leads to the proposition that modifications are needed to the institutional framework that governs the industry, and to the application and content of construction management training methodologies and delivery systems. In particular, the framework and systems should take proper account of the special needs of the target group of intermediate-level construction enterprises. These arguments lead in turn to the issue of "achieving sustainable change", which requires systems and procedures to enable these enterprises to develop and reach a sustainable level.

Appropriate construction technologies

The special problems and needs of developing countries demand the choice, transfer and application of special (appropriate) technologies and management techniques. The gradual acceptance of the concept of appropriate technology has had a considerable impact on the construction sector, since it provides a foundation for the definition of principles that should be applied in order to realise the potential for making more effective use of local resources.[1] There are five ways in which modern technologies can be modified so that they contribute effectively to development: by enabling the use of otherwise unusable local materials or resources, where the equipment and labour elements of the technology are easily separable, by replacing scarce skills, employing technical knowledge with little capital input, and by making an existing technology more effective.

Despite the economic and social attractions of the concept of appropriate technology, intermediate, labour-based technologies have not achieved wide acceptance except in the case of the construction industry (particularly road construction and maintenance). In relation to marketing and dissemination, the two most serious constraints to the choice and application of appropriate technologies are the fragmented market for intermediate technologies, which has inhibited research and development; and the lack of linkages and co-operation between the various organisations involved.

A further issue is the definition of policies for technology development and application, which requires an understanding of how technologies are developed, chosen and applied, based on an assessment of prevailing conditions in terms of the key variables of labour, capital, markets and resources. Policies are required to improve technology dissemination and implementation through attention to national policies and plans, support to national centres, improved communication, and small business promotion at five levels:

- improving technology choice;
- education and training curricula;
- professional training;
- vocational training; and
- management and supervisory training.[2]

The principles are illustrated by the experience of the ILO in promoting labour-based road construction and maintenance technologies, which offers a number of general lessons for engineers involved in international technology transfer including:

- the importance of assessing the local administrative, social, cultural and regulatory environment;

- the need for a sensitive assessment of the "software" factors governing the host organisation (skills, knowledge, experience, together with suitable organisational and institutional arrangements);

- the need for an open-minded approach to the choice of technology, taking account of national priorities in employment creation and the use of local resources;

- the importance of the principle of sustainability;

- the inseparability of technical and managerial innovation;

- the need to work with and through the local private sector, including initiatives to create a favourable contractual and regulatory environment; and

- the need to provide an appropriate mix of technical and management training to small construction enterprises, including cost accounting, estimating and bidding, and understanding contract documentation such as design specifications, drawings and payment procedures.[3]

A suitable regulatory framework

The framework governing the construction industry in industrialised countries (which has gradually evolved to suit the needs of the various actors) requires modification if the construction sector in developing countries is to be balanced and sustainable. Many developing countries modelled their regulatory framework on that which obtains in the United Kingdom. Thus UK experience can usefully be examined, with a view to gaining an insight into how accelerated development might be pursued in developing countries. A study of the way in which the UK construction industry developed gradually from its beginnings in the industrial revolution, showed how the sector developed its own checks and balances over a century of gradual evolution.[4] This study noted that the traditionally confrontational nature of the industry was already attracting increasing criticism in the mid 1970s, leading to subsequent interest in alternative procurement methods such as project partnering. It also helps to explain the difficulties encountered by those developing countries that came to independence with this system as an operational model, and the need for

163

modifications to permit and encourage indigenous enterprises to emerge. The UK model was sufficiently flexible to adapt to suit its own operational environment, but its transfer to the very different environment of a developing country may inhibit the growth of an effective local construction industry.

Special features

Two successive national studies of the Sudanese construction industry illustrate the special features of the sector in developing countries, the importance of the intermediate contracting sector and the scope for measures to strengthen it, together with the special constraints that need to be tackled in order to improve overall construction industry performance. In 1976 a large multi-disciplinary ILO "Comprehensive Employment Strategy Team" for the Sudan focused mainly on employment and productivity issues.[5] The chapter on the construction industry within the overall report identified significant weaknesses, including excessive reliance on casual labour, fluctuating and fragmented demand, problems in the manufacture and distribution of building materials and inadequate attention to repairs and maintenance, and drew attention to problems arising from the general inadequacy of construction management skills, covering the key resources of business organisation and finance, estimating and tendering, labour, materials and plant and equipment.

Five years later, a further study of the Sudanese construction industry was undertaken on behalf of the World Bank/UNIDO Co-operative Programme.[6] The purpose was to identify institutional constraints on sectoral performance, with a view to proposing a series of measures to improve the regulatory framework and the competitiveness of the domestic construction industry. It commenced with a review of construction demand and economic contribution, noting the paucity of reliable base statistics, and then described the organisation of the industry, including the role of public sector firms, foreign contractors and joint ventures as well as domestic construction enterprises. The latter group were disadvantaged in various ways, and the report recommended initiatives to strengthen the local contractors' association, improve payment procedures, provide investment incentives, revise contract conditions and procedures, and introduce an effective system of contractor classification.

National construction industry studies

International organisations, such as the World Bank, the ILO and UNIDO, used to take the lead in commissioning studies to investigate constraints on international construction industry performance. They are now less willing (or able) to support broad national construction industry studies, and instead proceed direct to detailed project identification and preparation. The lack of sponsorship for longer term, conceptual research is unfortunate, and a revival of interest in these issues would generally strengthen the foundations for future construction industry research. Should this revival occur, the analytical process and the analytical tools developed by Michael E. Porter offer a means to gain a better understanding of the nature of competition in the global construction market, including[7]:

- business strategy;
- defining the industry;
- determining industry structure;
- deciding positioning (including competitive advantage and competitive scope);
- analysing operations;
- analysing determinants of national advantage; and
- the sectoral features of engineering and construction.

Construction management training needs

Construction management training needs in developing countries differ from those in the industrialised countries where the discipline has evolved, and it is necessary to define target groups and develop suitable training methodologies and delivery systems in order to promote business and industrial efficiencies. The overview of construction management training needs in Chapter 4, defined target groups to be trained, covering policy-makers, construction programme managers, managers of construction operations and trainers, with special emphasis on small domestic (local) contractors.[8] The priorities that were identified in that overview still seem broadly applicable:

- The main target groups to be trained are policy-makers, construction programme managers, managers of construction operations and trainers, with special emphasis on small domestic (local) contractors.

- There is a need for attention to training needs, training methodology and delivery systems, emphasising the special needs of the various categories.

- Problems in developing countries include fluctuating demand, the casual nature of employment, choice of technology and the fragmentation of the industry.

- Research priorities are training needs, delivery of training and the respective roles of governments and employers' organisations.

Intermediate contractors

The intermediate level of construction, which lies between the traditional activities of rural builders and the more advanced technology of modern enterprises, is particularly important in developing countries. This activity is poorly understood, so the practices, problems and needs of intermediate contractors should be reviewed in order to offer appropriate training and development opportunities.[9] Studies of the practices, problems and needs of small contractors have established the link between the appreciation of the potential benefits of appropriate technology and the value of promoting small enterprises, which are best placed to apply it. The project experience suggested the following training priorities for intermediate contractors:[10]

- The business management of a contracting firm including marketing, book keeping and accounting, estimating and tendering, cost control, budgeting and forward planning.

- The operational management of building projects including contract procedure, site layout and organisation, job programming, plant management and maintenance, purchasing and storing of materials.

- Personnel and supervisory management with particular emphasis on intermediate technologies and labour intensive techniques.

A further study in 1993 described the origins and growth of the ILO Construction Management Programme, including the development of methodologies for small contractor development, and suggested ways in which the concept could be applied in developing and transitional economies.[11] Project experience led to an improved understanding of the importance of an emphasis on institution building in the design of technical cooperation projects, as well as tackling the policy constraints resulting from an inadequate regulatory and contractual framework. It concluded that the key factors in improving the performance of small contractors are:

- attention to appropriate training materials, coupled with a methodology for delivering linked training courses and on-the-job advisory services;

- understanding the importance of institution building in the design of technical cooperation projects; and

- tackling the policy constraints resulting from an inadequate regulatory and contractual framework.

6.2 Finance and funding

The long term objective of road funding must be for the recurrent and capital costs of the road network to be paid for from domestic sources. Full achievement of this objective is unrealistic in the short to medium term, and the majority of new road and other infrastructure provision depends directly or indirectly on donor finance. The key issue is to ensure that there is sufficient funding to maintain the current road network. This funding may be complemented by external assistance for rehabilitation work providing the domestic financial resources remains capable of paying the maintenance budget. External funding may also be provided as the seed 'capital' for road projects which provide the framework to ensure maintenance can be carried out.

It is clear that developing countries need to increase the amount that they spend on road maintenance - but by how much. Edmonds and de Veen argue that the 0.5% of GNP spent by industrialised countries would be unrealistic and that 0.35% GNP would be a more realistic figure for developing countries, and suggest that US $750 per km is required per annum to maintain the road network in a suitable state of repair.[12] Nevertheless this figure is often not

Table 6.1. Road networks in Africa and Asia					
	Actual length of road network	Potential length of road network		Length of road network with current levels of funding	
	km/1000pop.	km/1000pop.	% of actual length	km/1000pop.	% of actual length
Africa					
Burkina Faso	2.54	0.84	33	0.72	
Cameroon	6.77	3.88	57	4.79	71
Ethiopia	1.06	0.56	53	0.65	61
Guinea	2.41	1.4	58	0.78	28
Kenya	3.27	1.59	49	2.09	64
Malawi	2.32	0.98	42	0.89	38
Senegal	2.26	2.05	91	0.48	21
Asia					
Bangladesh	1.5	0.6	40	0.2	13
India	2.0	1.2	60		
Indonesia	1.0	2.6	260	0.37	37
Philippines	3.0	3.54	118	2.1	70
Sri Lanka	4.5	1.5	33	1	22
Thailand	3.0	3.8	127	1.7	57

Edmonds and de Veen, 1991

spent on road maintenance with the ultimate result that the road networks are moving to a poorer state of repair. At this level of investment they are able to show some sobering figures on the actual and potential size of road networks for a number of African and Asian countries. The columns in the table shows the actual length of the current road network compared with the potential length of the network for the spending described above and also compared with the current level of funding.

The table shows that an average of only 40% of the road networks for the countries listed above can be retained with the current level of spending on road maintenance. However, almost 80% of the road network can be retained if the level of spending on maintenance is increased to 0.35% of GNP. The figures above only include maintenance provision and no allowance has been included for rehabilitation of the road network that is currently in a poor state.

Mobilising domestic finance

The main conclusion to be drawn from the table above is that the size of road networks in developing countries have in general reached the maximum length that can be maintained by current domestic financial resources. Very critical consideration should therefore be given to programmes which propose the building of further roads until the domestic financial resources expand from their current level. Donor agencies interested in exploring the potential of supporting the development of the transport network in developing countries should also carefully consider the possibilities of assisting with rehabilitation and developing a sustainable maintenance system. Developing countries will need to increase their spending above the current level in order to maintain their networks. This additional domestic finance can come from three different levels:

Community level

It is unlikely that significant funds, usually in the form of unpaid labour, can be raised directly from the community themselves. Communities are generally unwilling to directly pay for the maintenance of their local network on a self help basis if others benefit from the work without making a contribution. Self help schemes invariably result in some communities undertaking more work than communities further from the road. In addition, at certain periods of the agricultural year there is unlikely to be free labour available to undertake roadwork.

Local Government

Traditionally funding for road maintenance has been channelled to local governments from the central ministry of finance. This funding has invariably been insufficient and erratic supply. If decentralisation of the road authority functions takes place, as suggested earlier, together with its associated financial autonomy there may be arguments in favour of raising finance for road maintenance at a local level. Potential sources of revenue that maybe raised at a local level include, property, sales tax and income however, none of these revenues are related to the provision of the road network. The only two potential local sources of finance from road use are road tolls and a local vehicle tax. Both of these revenue sources are very difficult to administer, with the potential for widespread evasion and hence limited financial returns.

Central Government

It is likely that in most cases the major source of funding for road maintenance will continue to come from a central government source collected through taxation. As maintenance budgets have invariably been only a fraction of the required amounts to maintain the integrity of the network, a system of providing a more realistic and reliable level of resources through some sort of road fund is highly desirable.

User charges

Traditionally roads have been financed through central funding, with little attention being paid to recovering the costs of providing this amenity from the direct beneficiaries. Unfortunately this means that road construction and maintenance has to compete for limited funds with education, social services and the full range of other government functions. This approach has resulted in the condition of the road network deteriorating as the other functions have vociferous supporters, and their needs appear more urgent. Developing country governments and donor agencies are becoming increasingly aware that they need to obtain a more sustainable source of funding in order to maintain road networks in their current condition (let alone improve them). The most common and widely accepted source of alternative funding for the road network is through direct charges of various kinds on road users. These charges are usually perceived as 'fair', since they put the burden of paying for the upkeep of the network onto the road users. They can take five basic forms; fuel levies, vehicle duties, road licences, road tolling and specific charges and fines.

Fuel levies

A fuel levy is probably the most effective method for raising finance to maintain the road network. The cost to the road user is proportional to the 'amount of road usage' and will be incurred at a steady rate through the year which will make the tax more acceptable to road users, particularly if they perceive that the money is being used to improve the road network. The fuel tax can be paid into a treasury account by importers when fuel is brought into the country and the additional cost passed on the consumer. The level of taxation in order to cover the costs of road maintenance by fuel levy has been estimated to be approximately 10 US cents per litre.[13] The cost of fuel varies widely across Africa with the highest cost of 98 cents /litre in Uganda to 13 cents per litre in Nigeria.[14] However, the average increase in fuel cost would be 15% across the African continent if the full cost of maintaining the road was raised through the fuel levy.

Vehicle duty

It would be very simple to charge a duty on the import of new vehicle with the duty being paid at the point of importation by the importer. This tax would then be passed on to the ultimate purchaser of the vehicle. Unfortunately this source of revenue will only provide modest funds if it is set at a realistic level compared with the original purchase price. If the import duty on the vehicle is about 10-20% of the purchase value, effectively a lump sum payment will be received when each vehicle is sold however, this lump sum will be significantly less than the revenue that would be generated by the same vehicle, during its whole life, from any of the other taxation systems. This method of obtaining resource to undertake road maintenance could only be used to supplement other revenue sources.

Road licences

A road licence is a simple means of collecting an annual fee for driving a vehicle on the road. Although it is relatively easy to set different rates according to the size or goods carrying capacity of the vehicle, it is difficult to administer the collection of the licence fee effectively. The common method is to require a licence disc to be displayed on the vehicle, but it is relatively easy to evade the licence fee in many countries, even if random road checks are carried out.

Road tolls

Road tolling is direct tax on the road users who will be required to make a payment for travelling along a particular section of road and hence it is often fiercely opposed. The collection of tolls is usually difficult as road users will take alternative routes, or may attempt to bribe toll collectors to allow them to pass without paying. Tolls at river crossings are usually easiest to collect, since alternative routes are inconvenient and costly. One system that seems to work well is the 'Octroi' tolling scheme which operates in some states of India. Tolls are collected on the main roads into towns with large payments being made for trucks and smaller payments for cars. There are no major alternative routes and the tolls are collected by a private company. The company receives a percentage share of the toll revenue which provides an incentive to effectively manage the collection process. The basis for charging trucks high tolls is that the importers of supplies into the town are profiting from the sale of these goods and are therefore best placed financially to meet the cost of maintaining the road network in their area.

Specific charges and fines

It is also possible to raise funds through specific charges, such as weighbridge charges, and fines for overloading heavy vehicles of other infractions such as speeding or using unroadworthy vehicles. The danger is that this is an unreliable source of funds for road maintenance, and tends to confuse the purpose of charges and fines (punishment of wrongdoers) with the purpose of road maintenance (providing safe and efficient access).

A road fund and a roads board

The roles of a road fund and a roads board were discussed in Chapter 3. One of the main tasks of the board for the road fund will be to allocate and distribute the funds from road maintenance. In many countries this will not be a simple task of passing the money to another department as the road network could be managed by a number of different organisations. For example, The Ministry of Works may be responsible for trunk or principal roads, the Ministry of Local Government responsible for district roads and local councils oversee the management of rural or village roads. The board will need to set clear criteria for the allocation of funds for which it is likely that a simple percentage-based distribution of the funds, as shown below, would be the most simple to administer and hence effective.

In the unlikely event that the board has additional funds above the minimum maintenance requirements it will be necessary to allocate funds in a more

Table 6.2. Typical distribution of road fund expenditure	
Type of road	*Allocation*
Principal/National	50%
Local/District	30%
Rural/Feeder	20%

complex manner probably through a bidding system. Each road management organisation would provide an annual budget which would outline essential maintenance costs (routine and periodic maintenance) and further work that it would like to undertake such as rehabilitation. It would then be necessary for the road board to review each bid and allocate the available funding according to their funding criteria. This system would be considerably harder to administer than the percentage distribution system and require closer auditing to ensure that the funds were distributed in a fair and effective manner.

Road Asset Management

The principle of treating roads as a portfolio of assets was introduced in Chapter 1. Regardless of the source of funding it will be necessary for the organisations managing road networks to decide how to utilise their financial resources so as to maintain and increase the value of this asset. As their needs will always be larger than the resources available it will be necessary to determine the most effective way to manage their maintenance programme. In these circumstances a road asset management system is likely to be the best method for managing the road maintenance budget.

Essentially the system works by assigning a value to each section of the road network (e.g. 1km or 5 km lengths) depending on its type of construction and condition of repair. It will then be possible to determine the total asset value of the network. Any maintenance or rehabilitation work undertaken will increase the value of that particular section of road by a given amount. If no work is undertaken on a section of road its value will decrease as it condition decreases. The budgetary objective is to spend the maintenance money to increase the total asset value of the road network by the maximum possible value. In the unfortunate case of the available financial resources not being able to increase the total asset value the objective will be to minimise the total decrease in value.

The first step is to define road condition, which can be in a few broad categories such as 'good', 'fair', 'poor' or 'lost'. In this case a 'lost road' is one which is impassable and/or overgrown. Even such a 'lost road' will still have a value as no major obstruction will exist, the ditches will still be constructed, although silted up, and the original profile may still be evident in places. The cost of reconstruction will therefore be cheaper than a new road. The road authority can then value its road network according to road lengths and condition, to produce a table valuing its portfolio of road assets as table 6.3 below.

Table 6.3. Typical road asset management system – value of roads (per km)					
Type	**Condition**				**Total**
	Good	*Fair*	*Poor*	*Lost*	
Regional bituminised	$??	$??	$??	$??	$??
Gravel	$??	$??	$??	$??	$??
Earth	$??	$??	$??	$??	$??
Total	**$??**	**$??**	**$??**	**$??**	**$??**

The authority will have work within an allocated budget, but the asset management system should help it to focus on longer term measures to strengthen the network and provide a better service to its clients. In practice it will have to decide how each section of road will increase in value if different levels of maintenance or rehabilitation are undertaken. It will also have to check how the value of each section of road will decrease if no work is undertaken. These calculations should highlight how the budget should be spent to result in the highest asset value of the road network at the end of the year.

Funding for contractor development projects

The discussion above has been concerned with the general financing of a road maintenance programme. It may be necessary to implement additional funding initiatives to assist contractor development schemes. These initiatives may ensure money is available on-time for contract payments or to assist contractors to purchase intermediate equipment or materials.

Revolving fund

A revolving fund may be used to assist contractors to purchase equipment or expand their business. The scheme requires seed capital to initiate the fund which will often come from donor support. When sufficient capital is available the fund will provide loans to contractors which meet the fund's criteria. Contractors will then be expected to pay back the loans and this money will be put back into the fund for future loans. As the scheme is not designed to make a profit, loans can be offered to contractors at a subsidised rate. The negative side of revolving funds are that they must be well managed to ensure that the capital is not lost on bad debts and that loans are granted for the intended purpose. An example of a revolving fund to enable contractors to be provided with a basic set of equipment is that operated by the Ghana Department of Feeder Roads (described in Chapter 5).

Use of local banks

Local banks can also be utilised in providing loans to contractors. The advantage of this initiative is that it shifts the administrative burden onto the banks which reduces the level of management required by the road authority. The interest rates charged by the banks will probably be higher than a revolving fund with soft (donor) finance as the banks will expect to make a profit. There must also be some method for reducing the banks' perceived risk of contracting as a business activity, in order that the interest rates charged to contractors are at a realistic level. This may be achieved either through bonds between the banks and road authorities to guarantee contractors are paid on time and/or payments to contractors being made directly into their bank accounts.

Special accounts

Apart from the problem of obtaining credit, the major financial problem experienced by contractors is late payments, particularly on projects that are financed through general taxation. Indeed, if clients were generally to make payments in strict accordance with the conditions of contract, much of the need for external borrowing by contractors would be obviated. One approach to delayed payments would be through the use of a special account, which is solely used to pay contractors' monthly certificates. Money could be deposited into the account from the road fund and additional funds provided from donor agencies. Contracts would not be let to contractors unless sufficient funds were in the account to pay the contractor.

6.3 Sustainable change

The performance of intermediate contractors can be significantly improved through appropriate guidance on technology and procedures using high quality material and delivery systems, providing attention is paid to sustainability based on an understanding of institutional and organisational change. This section seeks to draw together the various strands of thinking in the previous sections, in order to propose an approach to the development of intermediate contractors within a workable and sustainable market for construction services. It draws upon a variety of project experience in the promotion of the indigenous construction sector in developing countries, and emphasises the need to attend to attitudinal and procedural change in client organisations as well as directly promoting the emerging domestic contracting industry.

Contractor support agencies

An ICE technical note offers a tentative methodology for delivering technical assistance through contractor support agencies, which have links with both contractors and with government (as predominant client). [15] It explains that institutions of this kind are inherently fragile, and that it is not easy to maintain a basic autonomy and balance between links with contractors and clients in the face of contradictory pressures. The paper concludes with a discussion of ways in which agencies can be funded, with a mix between external support and direct user charges for items such as loan interest, bond and guarantee fees, course fees, consultancy and supervision charges, plant hire payments and charges for procurement of materials. It notes that an increasing element of direct financing is desirable since it provides a measure of the real demand for the agency's services. As direct charges are raised, non-economic services can be gradually abandoned so that the agency could eventually evolve into a self-sufficient institution run and financed by its beneficiaries.

Training methodologies and delivery systems

The growing world-wide emphasis on decentralised private sector provision of infrastructure projects and urban services in developing countries has resulted in a need to deliver training and other assistance to large numbers of nascent small enterprises. A study on the design and delivery of training and development programmes for small contractors in cross-cultural operational environments emphasised that the target group is hard to identify and harder still to reach, being dispersed over a variety of business activities in various sub-sectors, and based in a range of countries with different cultures and traditions.[16] Thus the trainer is faced with a dilemma. Some aspects of small enterprise development are common and widely replicable, which means that expenditure on developing high quality material and systems can be spread over numerous technical co-operation projects, but others are both sector and culture specific. The optimum solution is to prepare internationally applicable programmes and systems which can be supplemented with locally-based material to meet the special needs of particular sectors and national cultures.

Understanding the change process

Many developing countries lack a resourceful and experienced private sector which could readily be mobilised to meet new market opportunities. This implies a major change process, in which a local public sector monopoly supplier transforms itself into a commissioning and regulatory authority (or a number of such authorities), and a cohort of local contractors gradually emerges to undertake activities previously executed by direct labour. The cases in Chapter 5 show how the use of appropriate labour-based technology linked to the operation of market forces can generate significant savings through improved operational efficiency. However, in developing countries, the main difficulty is the lack of a resourceful and experienced private sector which could readily be mobilised to meet new market opportunities.

Delivery of technical assistance

The over-riding developmental consideration is *sustainability* - that is, the provision of technical assistance in such a way as to initiate beneficial change that will continue under its own momentum. Financing and aid agencies are accordingly turning away from supporting discrete infrastructure projects, towards the provision of multi-disciplinary consulting and advisory support, so as to ensure that their interventions have an enduring impact. Development projects are consequently expected to contribute to a broad technology transfer and skills development process, which may include courses and on-the-job training, but may also require 'twinning' and other linkages between institutions for a more substantial transfer of skills over a longer period.[17]

Project design and execution

A review of Asian experience in promoting private sector execution of construction projects and the use of labour-based methods drew on eight technical co-operation projects in countries with a variety of operational environments, distinguishing between projects where the task was contractor development and those where there was a more ambitious goal of construction industry development. This experience suggests five principles:[18]

- *Split responsibility:* Public sector as client/market regulator, service delivery by independent contractors.
- *Subsidiarity:* Decisions and execution delivered to level closest to consumer/beneficiary.
- *Sustainability:* All interventions should be designed to be sustainable within a properly regulated market.
- *Small enterprise focus:* Competitiveness is more likely to be sustained through promoting small enterprises.
- *System support:* Where small enterprises are weak, institutional intervention will be required.

6.4 Project evaluation

International financing and aid agencies involved in construction industry development expect their interventions to contribute to a broad technology transfer and skills development process, which may include courses and on-the-job training, but may also require 'twinning' and other linkages between institutions for a more substantial transfer of skills over a longer period.

Evaluating achievement is difficult since projects are so diverse in their technical demands, as well as in the range of target groups and circumstances involved. This section is illustrated by case studies from national technical cooperation projects based in China, Egypt, Ghana, India, Lesotho, the South Pacific and Sri Lanka, as well as regional and global initiatives. The lessons from the evaluation of these projects confirm the diversity of project experience, but suggest some common factors which could be taken account of at the project design stage to increase the likelihood of a successful outcome.

Implementing the change process

International technical assistance is an engine of development. It shares the characteristics of all engines; it can work well or badly - or even not at all, and it can be put to perverse uses. This paper is primarily concerned with the effective application of assistance to the task of institution building and organisational change within the construction/infrastructure sector, where the assistance is not merely technical (including the vital component of developing management and business skills), but must draw upon other disciplines and be delivered in a way which is sensitive to cross-cultural factors. It recognizes the essentially pragmatic approach to evaluation which international (as well as other) projects require, so that evaluation is, above all, about the task of understanding and then of explaining that understanding of the processes of change.[19]

A further consideration is that technical engineering skills are usually a prerequisite in designing, understanding and evaluating such projects, and those responsible need to combine an awareness of other disciplines with a knowledge of the special factors which characterise the construction sector. Growing interest in sectoral institution building and organisational change responds both to recipient countries' demand and the accumulating evidence from evaluation of past interventions, which have led financing and aid agencies to turn away from discrete infrastructure projects towards more ambitious programmes which will improve the performance of local industries to a stage where there is a realistic expectation that the improvements will sustainable without continuing external support.

Purpose of technical cooperation

The United Nations Development Programme (UNDP) defines the concept of technical cooperation as 'the establishment, strengthening or transformation of capacities in developing countries to plan, carry out, and assess their own development efforts.' Thus, 'most technical cooperation projects seek to develop human infrastructure in developing countries in contrast to (but often in conjunction with) capital projects that seek to develop the physical infra-structure. The two buzz words of development jargon - human resources development and transfer of technology - are usually subsumed in institution building: the human resources that are developed are always staff of the institutions being built up and the technology is usually, but not always, transferred to an institution as an element of its establishment, strengthening, or transformation'.[20]

Sustainability implies the provision of technical assistance in such a way as to initiate beneficial change that will continue under its own momentum, so the objective is not just to provide resources or to develop a set of defined skills, but to deliver an operational and sustainable system. In most cases, to be truly sustainable, this system should be capable of being operated and maintained in appropriate national institutions using resources that are locally available. Interventions should have an enduring impact by contributing to a broad technology transfer and skills development process, which may include courses and on-the-job training, but may also require 'twinning' and other linkages between institutions for a more substantial transfer of skills over a longer period. Indeed it can be argued that the goal of the development process is essentially to enable the beneficiaries to gain a more effective control of their environment.[21]

Characteristics of the construction sector

The construction industry is in general poorly documented and statistics relating to it are often unreliable, but there can be no doubt that it is very significant in virtually all developing countries both in economic and employ-ment terms. A study of 11 developing countries by the UNCHS suggested that its share of gross fixed capital formation is in the range of 35-83 per cent, that value added by the construction industry (excluding building materials and transportation) is typically in the range of 3-8 per cent and employment is in the range of 2-9 per cent (with a heavy clustering around 4-5 per cent).[22] Some of the features that characterise construction are the following; the responsi-bility for design is totally separated from the responsibility for production, the place of work constantly changes and is subject to interference from the weather, the work force comprises a large number of diverse specialised trades and employment is often casual in nature.

Construction relies on the successful deployment of a multitude of disparate technical skills, and it is also necessary to take account of local cultures, procedures, priorities and systems. In the context of public works in developing countries, sustainability is most likely to be achieved by helping people to help themselves, by means that fit naturally into their society and environment, so external consultants and contractors should develop the multi-disciplinary skills that are necessary to help communities to play a real role in decision-making.

Donors and recipient governments have also gradually become aware of the unusual scope for technological choice offered by the construction industry, and the resulting potential for maximising employment and the effective use of local resources.[23] An example is the World Bank's Sub-Saharan Road Maintenance Initiative, launched in 1989, which sought to promote both appropriate technologies and more effective use of the private sector through the use of local consultants and contractors.[24] Another strand of thinking relates to the role of development projects as part of a more general technology transfer and skills development process. This is usually through the provision of courses and on-the-job training, but may also be through linkages between institutions for a more substantial transfer of skills over a longer period.

A sustainable system

This discussion leads to the conclusion that the role of technical cooperation is not just to provide hardware in the form of infrastructure, but to deliver an operational and sustainable system. In most cases, to be truly sustainable, it should be capable of being operated and maintained using resources that are locally available.[25] This model is essentially:

DEPENDENCE > INTERDEPENDENCE > INDEPENDENCE

The problem is that there is a great diversity of construction work within the developmental, technical assistance framework. Although the principle of sustainability is now broadly accepted, its achievement in the construction sector is difficult since projects are so diverse in their technical demands, as well as in the range of target groups and circumstances involved. This diversity is illustrated by the following seven case studies from national technical cooperation projects, together with one regional and two global initiatives. The cases are introduced in table 6.4, which sets out briefly for each project the title, the duration, the main target group, and the most significant output. Following the case descriptions, the lessons from the projects are reviewed with a view to identifying key factors which favour sustainability in construction sector-related technical assistance programmes.

Table 6.4. Project evaluation case summary

Project	Duration (years)	Target group	Outputs
A. NATIONAL			
1. China: International construction management	4	Chinese construction corporations interested in project exports	Senior staff of large Chinese construction companies trained in modern financial and marketing techniques.
2. Egypt: Development of management information systems	2	Technical staff of large public sector contracting company	Computer-based management information systems developed for both site management and top management.
3. Ghana: Improve your Construction Business (IYCB)	3	Small-scale building contractors and building material manufacturers	Needs assessment, capacity to develop and deliver courses.
4. India: Management development in the construction industry	4	Academic institution (self-funding, non-governmental foundation)	Masters' programme developed, continuing capacity to deliver research and consultancy expertise.
5. Lesotho: Private sector maintenance of rural roads.	3	Prospective labour-based road contractors	Contractors selected and trained in both business and technical aspects.
6. Sri Lanka: Twinning arrangement to establish post-graduate construction management course	5	Academic institution	Construction management course established and supported by local industry.
7. Vanuatu (South Pacific): Cyclone Uma Reconstruction	1.5	Small-scale construction enterprises	Contractor development and promotion of labour-based technology.
B. REGIONAL			
8. Advisory Support, Information Services and Training (ASIST)	9	Project staff	National labour-based road projects in Eastern and Southern Africa supported.
C. GLOBAL			
9. Development of Construction Material Enterprises (DECO)	9	Small-scale manufacturers of building materials	Introduction of micro concrete roofing (MCR) tile technology
10. Management of Appropriate Road Technology (MART)	3	Road contractors and employers	Sectoral experience analysed and codified, results disseminated to practitioners, better linkages between relevant institutions and groups.

1. China International Contractors' Association

During the late 1980s large Chinese construction corporations began to compete aggressively in the international market for construction projects. Despite the comparative advantages of low costs and technical skills, they discovered that the risks in this market are commensurate with the potential rewards, and the China International Contractors' Association (CHINCA) was formed in 1988 to provide common services such as training and marketing information. They sought assistance from the ILO to secure appropriate business training for their middle-level and senior personnel, and the project output was defined as key staff trained to identify suitable markets, bid realistically and install planning and control procedures that would enable them to execute projects successfully.

The training covered a comprehensive range of topics that were further developed as a series of seven *International Construction Management* text books.[26] Due to heavy demand, 16 training workshops and seminars were delivered against 9 planned, and the number of direct participants was 870 against 450 planned. International experts were heavily involved in this pilot training, but the availability of the series of published text books enabled continuing dissemination as local institutions established their own training capability. Supporting positive evaluations from individual course participants, the growing confidence of Chinese construction corporations is indicated by the fact that the total volume of their construction and labour service activities in 1992 (the final year of the project) was US$ 6,585 millions, a 203 per cent increase compared to that in 1988, while four CHINCA members had figured in the list of the top 200 international contractors.[27]

2. Management information systems in Egypt

In the late 1980s, a large public sector contracting company operating within Egypt and neighbouring countries was faced with increasing competition as public construction enterprises were removed from the administrative control of sectoral ministries. Amongst other initiatives, it was urgent that effective management information systems should be developed, both to enable site management to improve the quality of short-term decision making and to enable higher levels of management to better assess current performance and business trends. The output was in the form of computer-based management information systems to provide:

- timely and relevant information to site management; and
- selective information to top management for more effective strategic control.

A technical assistance project was developed which included the supply of equipment and software packages, international fellowships and study tours for selected staff, consultancy advice from international specialists and training courses and workshops within Egypt. Sustainability is relatively easy to achieve in such a project, since it is possible to tailor inputs precisely to the needs of individuals in the light of their present and likely future roles within their organisation. The inclusion of fellowships and study tours enabled senior staff to establish useful linkages with possible partner institutions, and select those with which they would prefer to work. These personal contacts contributed to the success of the project, and the higher costs of tailoring training precisely to suit individual needs was justified by the urgent need to achieve higher levels of operational efficiency.

3. Ghana: Improve Your Construction Business (IYCB)

In Ghana, as in most developing countries, a relatively small number of major construction firms do a significant proportion of the available work, and a very large number of small firms do the rest. These small-scale contractors are a vital part of any developing economy, because they are available for all the multitude of small and domestic scale construction and maintenance work that serves the daily needs of a society. The stated output was that this group would be enabled to perform better by:

- determining their needs,
- developing suitable training material,
- delivering pilot courses in conjunction with a local institution and
- seeking to influence the business environment in which they operated so that it was more supportive of good business practice.

The main donor-financed inputs were an expatriate project manager together with provision for short term international and national specialist expertise, a project vehicle, equipment and limited travel funds. The choice of project manager was crucial. Fortunately it proved possible to recruit a candidate who had run his own small building business elsewhere in Africa before being recruited as a training adviser, and consequently had both training experience and a practical background with which the course participants could identify. The external assistance provided conceptual and general support, while an important element of the methodology was the direct involvement of local training specialists and contractors in the development and testing of the materials. The programme was pursued enthusiastically by the local contractors' association, which organised much of the field work, and found members to be trained as trainers. The decision to work with and through local institutions was essential in fostering local capacity to provide continuing inputs when external assistance ceased to be available.

The initial step was to determine the problems, practices and needs of the target group. This was done by circulating a questionnaire asking for a statement of problems and suggesting possible topics for training, supplemented by face-to-face discussions with individual participants at fact finding 'small-scale construction enterprise clinics'. In order to improve the prospect of lasting impact in Ghana and elsewhere, the training material was prepared and published as a series of handbooks and workbooks.[28] The final stage of project implementation was to initiate policy measures that would lead to the easing of regulatory and contractual constraints, and the project promoted a high level workshop to bring together the key actors. Although this did not lead to instant solutions, it did initiate a more constructive dialogue between the representatives of the contractors and the relevant Government departments contributed to the goal of enabling the beneficiaries to gain a more effective control of their environment.

The formal outputs promised by the Ghana project were significantly exceeded:

- 18 trainers trained (10/15 promised),
- 200 contractors trained (120/150 promised);
- three handbooks and three workbooks published (one of each promised).[29]

However, the new interest in contractors and their potential contribution to national prosperity could not be measured by statistics alone; CEBCAG and its members featured regularly in the local press and members of training cohorts were generally proud to be part of the national IYCB team. The IYCB symbol appeared on project publications and publicity material, and trainers carried IYCB brief cases and wore Improve Your Construction Business T-shirts! The costs of this marketing effort were modest, but were a significant factor in encouraging participants to take the training opportunities that were on offer and to feel that a neglected group of entrepreneurs were worthy of serious attention.

4. India: Management development in the construction industry

The Indian National Institute of Construction Management and Research (NICMAR) is a private charitable foundation, dedicated to research and postgraduate management studies to improve the performance of the construction industry. It arose in 1984 through an industry-led initiative, based on the recognition by an inter-ministry Task Force on Project Exports that "some form of management institute should be set up to cater for the specific needs of the construction industry" in order to improve the international competitiveness of Indian companies in the face of growing competition. When the technical assistance project was designed, NICMAR had been operational for just over a year, but was already running a one-year part-time evening course and a correspondence course for students working on remote sites. The technical assistance project had three immediate objectives:

- to upgrade the evening course to a two-year full-time Master's level programme;
- to consolidate and upgrade the correspondence course; and
- to develop research and consultancy capabilities.

The Director of the Institute, who was also designated as the National Project Director, had extensive previous international experience, which enabled him to build links with leading American, British and Canadian schools of construction management. The terminal review of this project concluded that its success could be traced to 'the establishment of an atmosphere of understanding and trust' between all parties to the project, the time taken at the start of the project to develop a clear implementation strategy (which led to a slow start, but which paid off in facilitating later progress), and fostering personal contact with international staff in partner institutions which continued well beyond the conclusion of the formal project.[30]

5. Lesotho: Routine Maintenance and Regravelling (ROMAR)

This project was described in some detail in the previous chapter. Essentially the Lesotho Government approached the ILO to facilitate the privatisation of public works activities, in order to generate savings through improved operational efficiency. The lack of indigenous road contractors meant that the first step was to select and train a cohort of local firms to undertake operational routine maintenance and regravelling (replacing the gravel wearing surface) of minor roads. The project benefitted from the use of the IYCB material (see Ghana case above) together with an additional package with the acronym ROMAR (for routine maintenance and regravelling). Two batches of contractors were selected and trained through a 12-month programme of classroom (business and technical) training and on-the-job training including test contracts.

The project was successful in its task of contractor development, with the target number of 15 accredited regravelling contractors attained and the target of routine maintenance contractors significantly exceeded (22 against a target of 15). However, an effective market requires an effective client organisation. It emerged that a weakness in the original project design was that insufficient attention had been given to achieving operational and attitudinal change among the senior and middle management staff responsible for contract award and supervision. Thus a key lesson of this project experience is the dual nature of the change process, and the consequent need to prepare the client organisation as well as the nascent contractors for their respective and complementary roles in the emerging market.

6. Sri Lanka: Twinning arrangement

Construction management has become an established discipline in the USA and Europe, where a Master's course in construction management is now seen as an important step for civil engineering graduates who intend to pursue a career in project management. Sri Lanka was one of the first Asian countries to appreciate the potential for postgraduate training in the discipline, and there appeared to be a sufficient potential demand since local produced around 180 civil engineering graduates per year. The quickest and most cost-effective way of instituting a national course seemed to be through a twinning arrangement with an established postgraduate programme. Loughborough University has two established MSc courses in Construction and Construction Management, which provided a framework, training material and a system which could be transferred and subsequently modified to suit local needs. The twinning arrangement allowed for junior staff from Moratuwa to undertake Loughborough MSc courses and a PhD programme, while Loughborough staff visited Moratuwa to support the programme in the early stages and to provide coaching assistance to other local staff.

The programme at Moratuwa became well established, and was underpinned by the University's strong local industrial contacts, which were deliberately fostered by providing seed funding to stimulate the creation of a market for contract research, as well as a growing international collaborative network. The confidence of the industry creates a virtuous circle, since potential students appreciate that the qualification is valued and past students demonstrate its value by their subsequent performance.

7. Vanuatu (South Pacific): Cyclone Uma Reconstruction

In February 1987 Cyclone Uma struck Vanuatu causing massive destruction, particularly on Efate and Tanna islands. In Port Villa, the capital, it was estimated that about 90 per cent of houses and 95 per cent of government buildings were affected. Devastation in the rural areas was also severe and many roads were damaged by landslides, flooding and washouts. Emergency assistance was pledged by several donors, but it was decided that the opportunity should be taken to develop long term capacity by promoting local small-scale contractors at the same time as introducing labour-based methods for road reconstruction and maintenance. The resulting 18-month project was successful in meeting a range of immediate objectives:

- construction supervisors and 24 labour-based road supervisors trained;
- courses on "small business management" provided for small-scale contractors;
- a Ni-Vanuatu Small-scale Contractors' Association was formed ;
- a register of contractors was established;
- improved rural housing techniques were introduced; and
- feeder roads constructed on Tanna and Efate islands.[31]

However, it is difficult to achieve long term change within the framework of a technical assistance project with specific short term objectives. An evaluation mission confirmed that the promised immediate objectives had been reached, but recommended that much more assistance was required to enable domestic contractors to become firmly established in their local market, since 'most of the domestic contractors remained small and fragile, with limited access to finance, and needing regular access to training and practical management advice - particularly in the areas of estimating and financial, commercial and site management'. Although donors were sympathetic, it proved impossible to secure resources for even limited follow-up activities to consolidate these achievements.

8. Regional: Labour-based roads in Africa (ASIST)

Interest in labour-based methods of construction in low-income countries has grown steadily over the past 30 years, and has been promoted mostly through individual field projects with intensive support from international specialists funded by international development agencies. Training is a key element of all these projects, and has to be provided at all levels from managers and supervisors to site personnel. The volume of activity in the Eastern and Southern Africa region suggested the need for external support, and a number of development agencies cooperated to establish a small regional support group under the auspices of the ILO, with the acronym ASIST (Advisory Support, Information Services and Training). ASIST is effectively a programme supporting a number of country projects, and also provides a Technical Enquiry Service and a clearing house for information exchange between projects in the region. The twin advantages of this programme are that 1) it provides support more economically than allocating expensive expatriate staff to a series of country projects and 2) the limited availability of ASIST staff for continuous 'hand holding' encourages national self reliance by ensuring that local staff enjoy real authority for operational decisions.

9. Global: Development of Construction Material Enterprises (DECO)

The inter-regional DECO project (DEvelopment of COnstruction Enterprises producing local building materials) started in July 1989 and was completed in 1998. It was executed by the ILO, but benefitted from association with an international network established by the Roofing Advisory Service of the Swiss Centre for Development Cooperation in Technology and Management. The objective was to demonstrate that the production of building materials is an important sub-sector of developing economies and that, when more appropriate technologies are promoted through the local private sector, their dissemination significantly to the generation of productive employment, an improvement of the balance of payments, and an increase of the incomes of small enterprises workers. The project sought to promote production, marketing and use of Micro Concrete Roofing (MCR) tiles in a number of African and Asian countries, as substitutes for imported materials (corrugated iron sheets) or heavily energy-consuming technologies (clay tiles). This entailed changes in all aspects of roof construction, including the construction of roofing structures.

The 9-year project was undertaken in three phases, each of about three years:

1. Introduction of the technology, to demonstrate the business potential for small enterprises of producing local MCR building materials in various countries.

2. Consolidation, to promote support mechanisms such as information centres and producers' associations, as well as direct promotion of enterprises which are viable and integrated into the construction market.

3. Disengagement, to consolidate the earlier interventions by strengthening local partners so that they would be viable when external support was no longer available. To take account of differing needs, it was necessary to define a disengagement strategy and criteria for each country.

The DECO project started as a technology promotion project, but developed into a private sector development project. It is notable for the long term commitment of the donor organisation, and the willingness of the executing agency to learn from experience in successive national interventions.

10. Global: Management of Appropriate Road Technologies (MART)

Many developing countries face deteriorating economic conditions, a crippling scarcity of foreign exchange with few assets other than an abundant supply of cheap labour. Efforts have consequently been directed towards developing and disseminating technologies which make more effective use of local resources (particularly human resources). Over the past twenty years, labour-based road construction technologies have been proved to be effective and economic in a wide variety of countries, and demand for advice and assistance on their implementation continues to grow. Since the focus has been on individual country projects, there was a need to draw together project experience and undertake generally relevant research on appropriate tools, equipment, training materials, documents and routines.

The Management of Appropriate Road Technologies (MART) initiative was based on a research project supported by the British Department for International Development (DFID), and was led by Loughborough University's Institute of Development Engineering in conjunction with specialist consultants Intech Associates and I.T.Transport. It aimed to encourage a wider application of this accumulated experience so that project interventions would not each have to retrace the same learning curve. Furthermore there was an emphasis on private sector involvement, so as to mobilise entrepreneurial skills and create enterprises which are be sufficiently flexible to provide lasting employment opportunities while reacting promptly to changing client requirements. The initiative sought to codify experience in the four key areas of handtools, intermediate equipment, private sector development and institution building.

The outputs are a wide range of published outputs (working papers, journal articles, handbooks and guidelines), which have been disseminated both directly and through workshops, seminars and conferences in a variety of developing countries. Practitioners have expressed the view that the MART project is notable for its collaborative and independent approach, which has enabled it to gather together and synthesise fragmented project experience from a wide variety of sources and disseminate best practice.

The lessons
The cases illustrate the diversity of development problems that beset the construction/infrastructure sector in most developing countries. Although the

Table 6.5. Key factors in achieving sustainability

Project	Outputs	Are outputs sustainable?	Outputs
A. NATIONAL			
1. China	Senior staff of large Chinese construction companies trained in modern financial and marketing techniques.	Yes	Experienced National Project Director worked closely with executing agency to guide implementation.
2. Egypt	Computer-based management information systems developed for both site management and top management.	Yes	Twinning links with experienced and compatible institution.
3. Ghana	Needs assessment, capacity to develop and deliver courses.	Yes	Project staff worked with and through local institutions.
4. India	Masters' programme developed, continuing capacity to deliver research and consultancy expertise.	Yes	Experienced National Project Director worked closely with executing agency and promoted industry links.
5. Lesotho	Contractors selected and trained in both business and technical aspects.	Partially	Insufficient attention to attitudinal change among client staff.
6. Sri Lanka	Construction management course established and supported by local industry.	Yes	Twinning links with experienced and compatible institution. Strong industry links.
7. Vanuatu	Contractor development and promotion of labour-based technology.	Partially	Insufficient project duration to consolidate improvements.
B. REGIONAL			
8. ASIST	National labour-based road projects in Eastern and Southern Africa supported.	Yes	Economical support to national projects and encouragement of self-reliance.
C. GLOBAL			
9. DECO	Introduction of micro concrete roofing (MCR) tile technology	Yes	Economical support to national initiatives and encouragement of self-reliance.
10. MART	Sectoral experience analysed and codified, results disseminated to practitioners, better linkages between relevant institutions and groups.	Yes	Independence from project executing agencies, collaboration with network partners and global reach.

promised outputs were in each case achieved, the project diversity makes it difficult to identify common factors which tend to lead to sustainable results and this is a factor in making sustainable sectoral development difficult to achieve. Those involved in project design must in each case define the target groups to be served, the types of external links to be forged, and the range of inputs and activities to be developed to serve them. The reported case experience is summarised in Table 6.5 above, focusing on the achievement of sustainability criteria rather than formal project outputs. The third column sets out the criteria for sustainability as described (or implied) in the project document, the fourth column assesses whether the success in meeting these criteria were met, and the final column comments on the reasons for achieving this degree of success.

Despite the diversity of case experience, Table 6.6 below suggests a number of common factors that tend towards success in achieving sustainable institution building in the construction/infrastructure sector. Positive experience is indicated with a tick (✔) and negative experience which confirms the proposition with a cross (✗).

6.5 The remaining challenge

This book has emphasised the importance of the intermediate level of construction in developing countries, and show how the practices, problems and needs of intermediate contractors in particular countries can be determined in order to formulate appropriate training and development opportunities. The key conclusion is that the performance of intermediate contractors in developing countries *can* be significantly improved providing the initiative takes proper account of constraints imposed by the local operating environment, provides a realistic plan for institutional and organisational change and uses high quality training material and delivery systems. This conclusion is of particular significance in view of the growing interest in promoting private sector involvement in infrastructure provision and maintenance. However, it also confirms that the task of construction industry development is inherently complex, and supports the view that

> ....the problems facing the construction industries of the developing countries are infinitely more fundamental, more serious and more complex, and their solution more pressing than those confronting their counterparts elsewhere.[32]

Table 6.6. Lessons learnt from case experience										
Proposition	**Case No.**									
	1	2	3	4	5	6	7	8	9	10
1. If a permanent project manager is required, he or she should possess excellent communication skills and work closely with local institutions.	✔		✔	✔	✔		✔			
2. When working towards comprehensive organisational change, foster attitudinal change as well as skills development in all parties.	✔	✔		✔	✗	✔				
3. Allow a sufficient project duration for assimilation and consolidation of the change process.				✔	✗	✔	✗	✔	✔	
4. Institution building can be reinforced through links with local industry and compatible overseas institutions.		✔		✔		✔	✔		✔	✔
5. There is a role for independent institutions and groups in providing common services and codifying and disseminating fragmented international project experience.								✔	✔	✔

To be appropriate to the priorities of construction industry development, research should recognise the special circumstances of developing countries, through comprehensive studies of national construction industries identifying factors such as productivity, costs, technology development, and so on, to provide the basis for policy formulation and corporate operating tactics. The lack of reliable data is a serious constraint on research on this topic, and field projects (which have the potential to generate valuable information) are only rarely designed with an integrated research and monitoring component to test out explicit and implicit assumptions. More work is required to generate additional information to analyse cross-cultural factors, so that the constraints on the performance of intermediate construction enterprises in each particular national business environment can be predicted and addressed.

References

[1]Miles, D.W.J. (1977) 'Selection, Transfer and Application of Appropriate Technology in Backward Regions'. Paper to UNIDO/TSKB Seminar on Industrial Projects Promotion in Istanbul. UNIDO, Vienna.

[2]Miles, D.W.J. (1982) *Appropriate Technology for Rural Development: The ITDG Experience*, ITDG Occasional Paper No.2, ITDG, London.

[3]Miles D. (1995) 'Choice and change: a case study' Chapter 2, pp 13-20 in D. Miles *Constructive Change: Managing International Technology Transfer*, International Construction Management Series No 5, ILO, Geneva.

[4]Miles, D. (1973) 'Development of the Construction Industry in Europe: United Kingdom Experience'. Paper to UNIDO Expert Meeting on the Construction Industry in Developing Countries. UNIDO, Vienna, .

[5]Miles, D.W.J. (1976) 'Construction' Chapter and Technical Paper on 'Organisation of the Construction Industry' in *Growth, Employment and Equity: A comprehensive Strategy for the Sudan*. ILO, Geneva.

[6]Miles, D. (1981) *Democratic Republic of the Sudan: Construction Industry Survey and Identification Report*. UNIDO/World Bank Co-operative Programme Report No.19 (id.82-24473), World Bank, Washington DC, USA.

[7]Miles, D. (1995) 'Understanding the Global Market' Chapter 1 pp 3-19 in D. Miles, *International Project Marketing,* International Construction Management Series No 6, International Labour Office, Geneva.

[8]Miles D. (1982) *Management Training for the Construction Industry in Developing Countries*. Report to the Tenth Session of the ILO Building, Civil Engineering and Public Works Committee, ILO, Geneva.

[9]Miles, D.W.J. (1998) *The Development of Intermediate Construction Enterprises*. PhD Thesis Loughborough University.

[10]Miles, D. (1973) 'Bi- and Multi-lateral Aid granted by the United Kingdom in the field of the Construction Industry'. Paper to UNIDO Expert Meeting on the Construction Industry in Developing Countries. UNIDO, Vienna.

[11]Miles D. (1993) *The Impact of the ILO Construction Management Programme on the Development of Small Construction Enterprises*, Construction Information Paper CIP/6, ILO, Geneva.

[12]Edmonds G.A. and de Veen J.J., (1991) *Technology choice for the construction and maintenance of roads in Developing Countries: Development and Guidelines*, (CTP 128), ILO, Geneva.

[13]Metschies, G., Rausch E., (1996) *Financing Road Maintenance*, Deutsche Gesellschaft fur Technische Zusammenarbeit (GTZ), Eschborn, Germany.

[14]Metschies, G. (1999) *Fuel Prices and Taxation*, Deutsche Gesellschaft fur Technische Zusammenarbeit (GTZ), Eschborn, Germany.

[15]Miles, D. (1983) 'Methodologies for the Delivery and Support to the Domestic Construction Industries in Developing Countries' (Technical Note 384); *Proceedings of the Institution of Civil Engineers Part I*, pp 74.

[16]Miles, D. (1995) 'Training Across Boundaries: Promoting International Small Enterprise Development'. *Cross-cultural Management*, Vol.2, No.3, pp 39-46.

[17]Miles, D. (1996) 'Effective Technical Co-operation for Construction Industry Development'. Paper to CIB Conference on Construction Modernization and Education, Beijing.

[18]Miles, D.W.J. (1997) 'A Decade of Small Contractor Development in Asia: Lessons from project experience', *Public Works Management and Policy*, Vol. 1, No.3, pp 245-257.

[19]Van der Eyken, W., Goulden D. and Crossley M. (1995) 'Evaluating Educational Reform in a Small State'. *Evaluation*, Vol 1, No 1, pp 42-43.

[20]UNDP (1990) *How to write a project document: A manual for designers of UNDP projects*. UNDP, New York.

[21]Miller, E. (1993) *From dependency to autonomy: Studies in organization and change*. Free Association Press, London.

[22]UNCHS (1984) *The construction industry in developing countries: Vol. 1. Contributions to socio-economic growth*. United Nations Centre for Human Settlements - UNCHS (Habitat), Nairobi.

[23]Edmonds, G.A. and de Veen J.J. (1992) 'A labour-based approach to roads and rural transport in developing countries'. *International Labour Review*, Vol 131, No 1, pp 95-110.

[24]Lantran, J.M. (1996) *Contracting Out Road Maintenance Activities: A World-wide Trend*. ASIST Bulletin No.5. ILO/ASIST, Nairobi.

[25]Miller, E. (1993) *op cit*.

[26]ILO (Various authors) (1995). *International Construction Management series*. ILO, Geneva.

[27]ILO (1993) 'UNDP/ILO Project CPR/88/024. Terminal Report by National Project Director'. ILO, Geneva (unpublished).

[28]Andersson, C-A, Miles, D.W.J., Neale, R.H. and Ward, J. (1994/6) *Improve Your Construction Business series*. ILO, Geneva.

[29]Miles, Derek (1993) *op cit*. p12.

[30]Miles, Derek. and Neale, Richard (1991) *Building for Tomorrow: International Experience in Construction Industry Development*. International Labour Office, Geneva.

[31]ILO (1991) 'Project VAN/87/011: Project Findings and Recommendations'. ILO, Geneva, (restricted).

[32]Ofori, G. (1993) 'Research on construction industry development at the crossroads'. *Construction Management and Economics*, Vol 11, pp 175-185.